Food Rh

T0074819

Rheology is the study of material flow and deformation, defining different aspects of food processing and product design. Food rheology affects several unit operations during food processing, the behavior during shelf life, the consumer perception during consumption and the interaction of food products with the human body (from chewing and swallowing to digestion). Therefore, it is imperative for professionals involved in food science and engineering to understand and assess food rheology.

Food Rheology: A Practical Guide presents the main aspects of food rheology as a practical guide, demonstrating that applying food rheology does not need to be a complex task.

Key Features

- Presents a practical, direct and didactic description of food rheology, with many examples and applications
- Includes a guide for designing, performing and interpreting experiments, highlighting the main concerns and tips
- Describes different food products (liquid, semi-solid and solid; homogeneous and heterogeneous; vegetable- and animal-based) with examples and applications
- Explores structure-processing-properties relations

More direct, practical and consulting, this book can help students, professionals and professors to understand the basic concepts to design, perform and interpret experiments related to food processing and properties.

Food Rheology
A Practical Guide

Pedro E. D. Augusto, Meliza L. Rojas
and Alberto C. Miano

CRC Press
Taylor & Francis Group
Boca Raton London New York

CRC Press is an imprint of the
Taylor & Francis Group, an **informa** business

First edition published 2024
by CRC Press
2385 NW Executive Center Drive, Suite 320, Boca Raton FL 33431

and by CRC Press
4 Park Square, Milton Park, Abingdon, Oxon, OX14 4RN

CRC Press is an imprint of Taylor & Francis Group, LLC

© 2024 Pedro E. D. Augusto, Meliza L. Rojas, and Alberto C. Miano

ISBN: 978-0-367-70968-6 (hbk)
ISBN: 978-0-367-68569-0 (pbk)
ISBN: 978-1-003-14872-2 (ebk)

DOI: 10.1201/9781003148722

Typeset in Garamond
by Apex CoVantage, LLC

We would like to dedicate this book, acknowledging, to:

—our students, whose curiosity and interest
encourage us to keep the march,

—our professors, whose wisdom and
teaching allowed us to get here,

—our family, whose unconditional support
and love give reason to the journey.

"If I have seen further, it is by standing on
the shoulders of giants."

Isaac Newton

Contents

Preface

Rheology is the study of material flow and deformation, defining different aspects of food process and product design. Food rheology affects several unit operations during food processing, the behavior during shelf life, the consumer perception during consumption and the interaction of food products with the human body (from chewing and swallowing to digestion). Therefore, it is imperative for professionals involved in food science and engineering to understand and assess food rheology.

Pedro E. D. Augusto is professor at the Université Paris-Saclay, CentraleSupélec, France. Alberto C. Miano and Meliza L. Rojas are professors at the Universidad Privada del Norte, Peru. However, this book started at the Universidade de São Paulo, Brazil, where the authors worked together for some years.

This book was born as a demand by the authors. By developing and evaluating new food processes, the authors missed a practical guide of food rheology.

Therefore, this book presents the main aspects of food rheology as a practical guide. It does not intend to substitute the good and classical books of food rheology we can find in the literature, but instead to be a more direct and practical book for students, professionals and professors to consult to help them in their daily efforts.

In addition, although this book does not intend to be exhaustive, it can be a friendly and bedside book, helping you to perform experiments and interpret the results related to food processing and properties.

We hope that you will find this book practical and helpful, and we would love to hear your opinion about that.

Good work!

The authors

About the Authors

Pedro Esteves Duarte Augusto is a full professor at Université Paris-Saclay (France), CentraleSupélec, being the vice-director of the chair of biotechnology. He works in the Centre Européen de Biotechnologie et de Bioéconomie (CEBB), an advanced research campus in Pomacle, studying food and bioproducts processing, including process engineering and biomaterials. He is a former professor of the University of São Paulo (USP) and University of Campinas (UNICAMP), both in Brazil. Prof. Augusto has an habilitation degree (HDR) in food processing from the University of São Paulo, postdoctorate as Visiting Professor at the University of Lleida (Spain), Ph.D. in food technology from the University of Campinas, with a period at the University of Lleida, and M.Sc. in food technology, food engineering and food technician from the University of Campinas.

Alberto Claudio Miano Pastor is Head of the Advanced Research Center in Agro-engineering at Universidad Privada del Norte (UPN, Peru), working with food processing engineering. He obtained his Ph.D. and M.Sc degrees in food science and technology from the University of São Paulo (USP, Brazil). Prof. Miano has experience in food processing as mass transfer, hydration, drying, acidification, food structure and functionality, solid viscoelasticity, texture analysis and emerging technologies.

 **Meliza Lindsay Rojas Silva** is a research professor at the Universidad Privada del Norte (UPN, Peru) working with food technology and processing. She obtained her Ph.D. and M.Sc degrees in food science and technology from the University of São Paulo (USP, Brazil), with a period at the Polytechnic University of Valencia (UPV, Spain). She is also an agroindustrial engineer (UCV, Peru). Prof. Rojas has experience in emerging technologies in food processing, thermal processing, kinetics of reactions, mathematical modeling, drying, food structure and functionality and physical and nutritional properties, including rheology.

Introduction and Basic Concepts of Rheology

Alberto C. Miano, Meliza L. Rojas and Pedro E. D. Augusto

1.1 Rheology Background

The consumer demand for new food products is growing, especially for products which are more like natural raw materials, with reduced environmental impact, with new sensorial properties or with better nutritional quality. This creates the need for a better understanding of food processing and properties. Indeed, one technique used for these is by rheological analysis, which is important for the design of unit operations, process optimization and high-quality product assurance. Rheology is the science that studies the flow and deformations of solids and fluids under the influence of mechanical forces. This is used not only for food processing but also for many other fields of study such as geology, concrete technology and polymers, among others (Rao, 2013).

1.2 Rheology Importance

The rheological properties of foods are used to design equipment and processing plants, as well as food packages and even to estimate shelf life. In addition, rheology has started to be used to understand the interaction food with the human body. It dictates the transport phenomenon to and through the product, comprising the mass transference (as the diffusion of ingredients during processing or sedimentation during storage), momentum (as the velocity profile in pipes or the propellers requirement during mixing) and energy (as the heating rate in heat exchangers and vessels). Therefore, not only are process conditions and dimensions (such as pump power, pipes, evaporators and heat exchanger dimensions or temperature profile during thermal process) defined based on the product rheology, but also the product behavior during storage (as the separation of phase or gelation). Rheological properties are also important for food quality control, being used to guarantee desirable standards. Finally, rheology is directly related to the consumer's sensorial perception, being essential for its acceptance.

DOI: 10.1201/9781003148722-1

1

For instance, fruit and vegetable juices are composed of an insoluble phase (the pulp) dispersed in a viscous solution (the serum). The dispersed phase, or pulp, is constituted of fruit tissue cells and their fragments, cell walls and insoluble polymer clusters and chains. The serum is an aqueous solution of soluble polysaccharides, sugars, salts and acids. The fruit juice rheological properties are thus defined by the interactions within each phase and between them (Augusto and Vitali, 2014). Therefore, the fruit juices show complex rheology, described by non-Newtonian behavior, with time-dependent, shear-thinning and viscoelastic (either viscous or solid) properties.

Another example is with some dairy products. During yogurt processing, many reactions occur such as acidification, change on protein stability, calcium solubility and complexation, protein denaturation, and protein-fat interactions. These cause changes in rheological properties of yogurt during processing time. Therefore, if processing parameters are not correctly controlled, the rheological quality of the final product would be undesirable. Furthermore, by analyzing rheological properties, new technologies to improve the process can be studied and used to determine if some effects on the yogurt consistency occur (Oliveira et al., 2014).

Therefore, food rheology is function of its composition, as well as due to processing and its condition. Moreover, the study of food rheology—as well as the effect of products and processing conditions on its properties—is fundamental to guarantee high-quality products.

1.3 Fundamentals of Material Properties

1.3.1 Stress–Strain

Rheology analyzes the material reaction due to a stress–strain assay. In other words, it studies the material deformation due to a stress or the forces generated when a material is released from a deformation. From this, two concepts are important for further understanding:

1. Stress (σ), which is the force (F) per area unit (A) which is applied to a material (Equation 1.1).

$$\sigma = \frac{F}{A} \tag{1.1}$$

There are different ways of applying a force to a material, and depending on their orientation, they get different names (Figure 1.1). When the force is exerted oriented to the sample, perpendicular to the surface

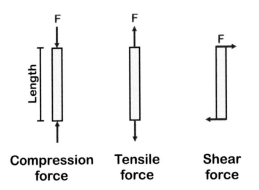

Figure 1.1 *Types of applied forces depending on their orientation.*

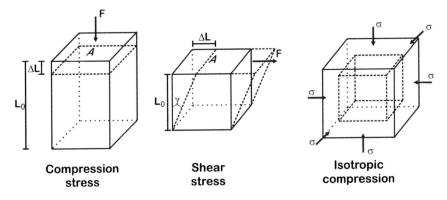

Figure 1.2 *Types of stresses depending on their applied orientation.*

applied, it is known as uniaxial compression force. On the other hand, if the force is oriented to pulling the sample, it is known as tensile force. In addition, when a force is applied tangentially to the surface, it is known as shear force.

Relating these forces to a specific area, different types of stresses can be obtained (Figure 1.2): uniaxial compression, when a simple compression in one direction is applied; bulk or isotropic compression, when stress is applied to all directions (hydrostatic pressure, for instance); and shear stress, when a force is tangentially applied to one area, locking the lower area.

2. Strain (ε), which refers to the change of material's dimensions caused by a stress. There are two ways for calculating strain: Cauchy strain (engineering strain) (Equation 1.2) and Hencky strain (true strain) (Equation 1.3).

$$\varepsilon_c = \frac{\Delta L}{L_0} \tag{1.2}$$

$$\varepsilon_h = \int_{L_0}^{L} \frac{dL}{L} = \ln\frac{L}{L_0} \tag{1.3}$$

Where L_0 is the length of the original sample (unstressed), L is the length of the stressed sample and ΔL is the length variation ($|L-L_0|$).

Another used strain in rheology is the one caused by a shear force (Figure 1.2). This force forms a deformation angle (γ) which can be calculated using Equation 1.4.

$$\tan(\gamma) = \frac{\Delta L}{L_0} \tag{1.4}$$

Particularly, when small deformations occur, the angle of shear (in radians) can be considered equal to the shear strain (Equation 1.5).

$$\tan(\gamma) \sim \gamma \tag{1.5}$$

In general terms, nine separate quantities are required to completely describe the stress state of a material: solid or fluid. Working with cartesian coordinates, stress is quantified by σ_{ij}, where "i" subscript is related to the orientation of the face area upon which the force acts; and "j" subscript indicates the direction of the force (Figure 1.3). These nine stress components can be summarized in a matrix known as stress tensor (Equation 1.6).

$$\sigma_{ij} = \begin{bmatrix} \sigma_{xx} & \sigma_{xy} & \sigma_{xz} \\ \sigma_{yx} & \sigma_{yy} & \sigma_{yz} \\ \sigma_{zx} & \sigma_{zy} & \sigma_{zz} \end{bmatrix} \tag{1.6}$$

Further, we can identify normal stresses and shear stresses in Figure 1.3. Normal stress is the stress whose force is perpendicular to the applied area. This means that σ_{ij}, whose subscripts are the same coordinates, are normal stresses, such as σ_{xx}, σ_{yy} or σ_{zz}. On the other hand, shear stress is the one whose force is parallel to the applied areas. This means σ_{ij} whose subscripts are different, such as σ_{xy}, σ_{yz} or σ_{zx}.

From this information, any equation that shows the relation between stress and strain is called rheological equation of state or constitutive equation. This could be for solids or fluids.

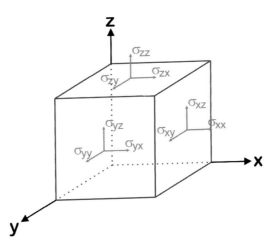

Figure 1.3 *Stress state of a material: solid and fluid. Stresses are represented by σ_{ij}, where "i" subscript is related to the orientation of the face area upon which the force acts and "j" subscript indicates the direction of the force.*

1.3.2 Solid Behavior

When stress–strain studies (stress vs. strain) are conducted for ideal solids, they generate a line that crosses the origin in cartesian coordinates. This behavior is characterized by elastic solids, and it is known as Hooke's Law (Equation 1.7), where the slope is represented by the Young's modulus of elasticity (E).

$$\sigma_{ij} = E \cdot \varepsilon \tag{1.7}$$

Hookean materials are characterized by having linear elasticity. This means that they recover their dimensions after stress release. However, many food materials present this behavior only by using small strains (less than 0.01)—for example, dried pasta and candies—otherwise, it generates material fracture or non-linear behaviors. For more detail, see Chapter 3 and Chapter 6.

This behavior can be studied by uniaxial compression test for some common geometries. For instance, Figure 1.4 shows a compression test for a cylinder shape sample. If deformation is small enough to depreciate radius variation, Equation 1.8 can be used for calculating elasticity modulus (E); otherwise, radius variation must be considered, and Equation 1.9 should be used.

$$E = \frac{stress}{strain} = \frac{\sigma}{\varepsilon_c} \tag{1.8}$$

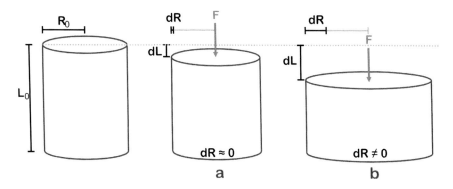

Figure 1.4 *Examples of compression tests. a. When deformation is small for depreciating radius variation. b. When deformation is big enough to consider radius variation.*

$$E = \frac{F}{\pi \cdot (R_0 + dR)^2} \tag{1.9}$$

Another useful test to determine elasticity modulus is the flexural test of cantilevers (Figure 1.5). This test consists of applying a force in order to deflect a cantilever maintaining one or more fixed points (in general, two). It should be considered that calculation depends on cross-section geometry. In fact, it can be used for food similar to cantilever shapes as spaghetti, cheese snacks, bakery products and so on.

Regarding the simple flexural test, Equation 1.10 and Equation 1.11 are used for rectangular and circular cross section, respectively. For three-point bending test, Equation 1.12 and Equation 1.13 are deduced for rectangular and circular cross sections, respectively.

$$E = \frac{4 \cdot F \cdot a^3}{d \cdot b \cdot h^3} \tag{1.10}$$

$$E = \frac{64 \cdot F \cdot a^3}{3 \cdot d \cdot \pi \cdot D^4} \tag{1.11}$$

$$E = \frac{F \cdot a^3}{4 \cdot d \cdot b \cdot h^3} \tag{1.12}$$

$$E = \frac{4 \cdot F \cdot a^3}{3 \cdot d \cdot \pi \cdot D^4} \tag{1.13}$$

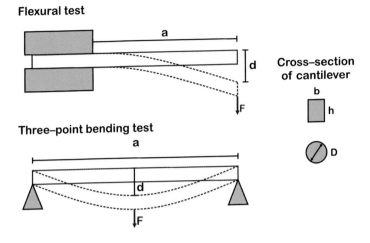

Figure 1.5 *Common examples of flexural tests of cantilevers.*

Since food materials do not behave as a perfect elastic solid due to their complex structures, Young's elasticity modulus should be carefully used. Therefore, some researchers recommended using Apparent Young's Modulus or Modulus of Deformability (Bourne, 2002; Kramer and Szczesniak, 1973). Some values of these parameters are 0.1–0.3 Pa for bread, 8–30 Pa for fresh banana and 200–400 Pa for raw carrot.

Other important moduli are the Shear Modulus and the Bulk Modulus. Shear Modulus is also called modulus of rigidity, and is the relation between shear stress (τ) and shear strain (Equation 1.14). On the other hand, Bulk Modulus is the relation between isotropic stress (hydrostatic pressure, P) and volumetric strain (Equation 1.15).

$$G = \frac{shearing\ stress}{shearing\ strain} = \frac{\tau}{\gamma/L_0} \qquad (1.14)$$

$$H = \frac{hydrostatic\ pressure}{volumetric\ strain} = \frac{P}{dV/V_0} \qquad (1.15)$$

As stated before, food products do not behave as Hookean solids (more details in Chapter 3). There are other types of behaviors, as shown in Figure 1.6. Elastic solids are characterized by recovering their initial dimensions when stress is released. This elasticity may be linear

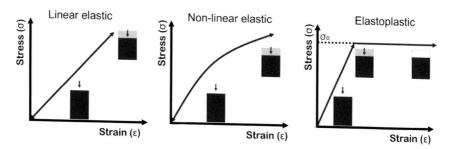

Figure 1.6 *Stress–strain behavior of solid food under compression.*

(Hookean), as iron, candies and dried pasta; or non-linear, as rubber. Further, other solids are elastoplastic, which behave as Hookean solids until certain yield stress (σ_0), then permanently deform. Margarine and butter at room temperature behave as elastoplastic solids.

1.4 Fluid Structure and Properties

Gas and liquid state substances are considered fluids. In contrast to solids, fluids differ by their capacity to react against a shear stress, which is a consequence of molecular mobility. Solid opposes resistance to a shear stress, but fluids deform continuously when a shear stress is applied. For instance, placing the substance between two parallel planes and moving upper plane keeping lower plane fixed, a shear strain is performed (Ibarz and Barbosa-Canovas, 2014)—see Figure 1.7 and more information in Chapter 2. Therefore, a deformation occurs creating a shear angle (α), by moving points a and b to a' and b'. When the stress is released in a solid, two common scenarios can happen: the solid will recover its initial dimensions (if it is an ideal, elastic solid) or the solid will maintain its deformation permanently (if it is a plastic solid). Contrary to this, if the same stress is applied to an ideal fluid, it will keep deforming even if the stress is released.

Similar to a solid, fluids are affected by different forces. Figure 1.8 shows the forces that affect a control volume of a liquid, like a solid block on an inclined plane (Penaloza-Giraldo, 2019). This control volume has a mass that generates weight force (F_W), which is divided in one for each coordinate axis (two forces for two dimensions coordinates, $F_{W,x}$ and $F_{W,y}$.). In y-axis, $F_{W,y}$ is balanced by the normal force (F_N). On the other hand, in x-axis, $F_{W,x}$ is equal to friction force (F_F) if the volume control is static, or is accelerated if the volume control is accelerated (according to Newton's laws). Normal force and friction force (shear force is used for fluids instead of friction force) are considered surface forces since act across internal or external surface. Further, weight force is a body force since acts throughout the volume of a body.

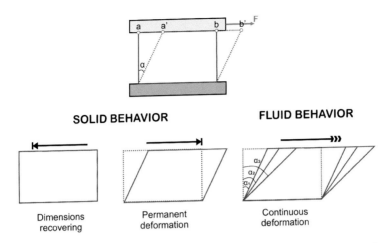

SOLID BEHAVIOR

FLUID BEHAVIOR

Dimensions
recovering

Permanent
deformation

Continuous
deformation

Figure 1.7 *Difference between solid and fluid behavior against a shear force.*

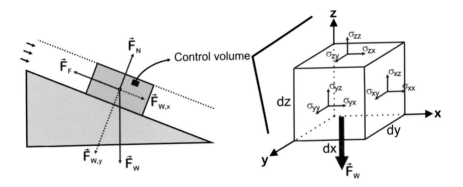

Figure 1.8 *Forces and stresses that affect a control volume of a fluid.*

The relation between surface force and its respective surface can be as normal stress and shear stress (Figures 1.1 and 1.2). These stresses are also known as viscous stresses, since they cause resistance for fluid flowing. Equation 1.16 is an example of normal stress from Figure 1.8 control volume, and Equation 1.17 is an example of shear stress from Figure 1.8 control volume.

$$\sigma_{xx} = \frac{F_{xx}}{dxdy} \tag{1.16}$$

$$\sigma_{xy} = \frac{F_{xy}}{dxdz} \tag{1.17}$$

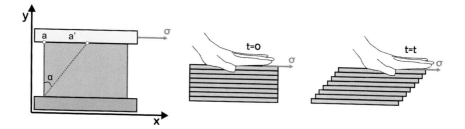

Figure 1.9 *Schematic representation of velocity gradient generated in a fluid under shear stress.*

Viscous stresses prevent fluids from flowing constantly. This fluid property is called viscosity, which is very important in rheology and its determination is crucial for many processes (see Chapter 2 and Chapter 5). Viscosity is defined as the fluid property that opposes to movement and deformation. This property is caused by the interaction of the fluid molecules. Going back to the two planes experiment: a shear stress is applied to the upper plane, keeping lower plane fixed. This caused a displacement (x) of the fluid as in Figure 1.9, from point a to point a', forming a shear angle α. To get a better idea of the experiment, imagine placing the palm of your hand on top of a deck of cards and moving horizontally. You will notice that the top cards move at the same speed as your hand, while the farthest cards move less (or even do not move, staying at rest as the table where they are leaning). Similarly, displacement behavior happens on a fluid. In fact, the movement performed in the uppers plane is transferred by the fluid particles or molecules to the lower plane. However, this displacement is reduced due to energy dissipation by friction—in other words, due to viscosity. In fact, this process is not always linear, as described in Chapter 2.

Now, considering an instant of time, each displacement $(x_1, x_2, x_3 \ldots)$ can be divided by time (t) and represented by velocities $(u_1, u_2, u_3, \ldots)$ (Equation 1.18) (Figure 1.10).

$$\frac{u_{max}}{y_{max}} = \frac{u_1}{y_1} = \frac{u_2}{y_2} = \frac{u_3}{y_3} = \frac{u_4}{y_4} = \frac{du}{dy} \qquad (1.18)$$

On the other hand, as the deformation time increases, the displacement and shear angle are also increased (Figure 1.10). Using tangent concept, the differential of the shear angle can be estimated and expressed as the relation of the maximum displacement and maximum plane separation (Equation 1.19). In addition, the maximum displacement velocity is

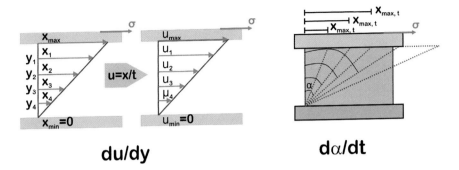

du/dy $d\alpha/dt$

Figure 1.10 *Fluid deformation due to a shear stress. At an instant of time, a velocity gradient is formed. On the other hand, the shear angle is increasing as the time passes.*

defined as the ratio between the maximum displacement and the time (Equation 1.20).

$$d\alpha \approx tan\alpha = \frac{dx_{max}}{y_{max}} \qquad (1.19)$$

$$\frac{dx_{max}}{dt} = u_{max} \qquad (1.20)$$

Replacing Equation 1.18 and 1.20 in Equation 1.19, Equation 1.21 can be obtained. This can be summarized as in Equation 1.22.

$$d\alpha \approx tan\alpha = \frac{dx_{max}}{y_{max}} = \frac{u_{max} \, dt}{y_{max}} = \frac{du}{dy} \, dt \qquad (1.21)$$

$$\frac{d\pm}{dt} = \frac{du}{dy} \qquad (1.22)$$

Indeed, the cause of the deformation expressed as shear angle variation or velocity gradient is the shear stress (σ). Consequently, the expression on Equation 1.22 is directly proportional to the shear stress. For convenience, it is more suitable to use the velocity gradient (Equation 1.23), instead of shear angle variation (Equation 1.24) because it is easier to measure in many situations.

$$\sigma \propto \frac{du}{dy} \qquad (1.23)$$

$$\sigma \propto \frac{d\alpha}{dt} \qquad (1.24)$$

Further, a coefficient should be added to Equation 1.23 to control the deformation and to characterize the fluid. This coefficient is the viscosity. Therefore, Equation 1.25 is obtained. To solve this equation, it should be integrated. However, if the velocity gradient has linear behavior, its solution can be simplified as Equation 1.26.

$$\sigma = \mu \frac{du}{dy} \tag{1.25}$$

$$\sigma = \mu \frac{u_{max}}{y_{max}} \tag{1.26}$$

Now, Equation 1.26 can be extended by inserting the definition of shear stress (Equation 1.1), which is the relation between the shear force and the applied area (Equation 1.27). This is helpful for viscosity measurement. In fact, geometries with known and constant areas are used to facilitates even more its measurement, obtaining Equation 1.29.

$$\frac{F}{dA} = \mu \frac{u_{max}}{y_{max}} \tag{1.27}$$

$$F = \mu \frac{u_{max}}{y_{max}} dA \tag{1.28}$$

$$F = \frac{1}{4} \frac{u_{max}}{y_{max}} A \tag{1.29}$$

Nomenclature

α = shear angle (Equations 1.19, 1.21, 1.22, 1.24) [-]
ε = deformation level or strain [-]
ε_c = engineering strain (Equation 1.2) [-]
ε_b = true strain (Equation 1.3) [-]
σ = stress (Equations 1.1, 1.23, 1.24, 1.25, 1.26) [Pa]
σ_{ij} = stress, where "i" subscript is related to the orientation of the face area upon which the force acts; and "j" subscript indicates the direction of the force (Equation 1.6, 1.16, 1.17) [Pa]
μ = viscosity (Equations 1.25, 1.26, 1.27, 1.28, 1.29) [Pa·s]
γ = deformation angle (Equation 1.4) [-]
A = area (Equations 1.1, 1.27, 1.28, 1.29) [m²]
b = width (Equations 1.10, 1.12) [m]
d = flexural deformation (Equations 1.10, 1.11, 1.12, 1.13) [m]
D = diameter (Equations 1.11, 1.13) [m]

E = modulus of elasticity (Equation 1.7) [Pa]

F = force (Equations 1.1, 1.10, 1.11, 1.12, 1.13, 1.27, 1.28, 1.29) [N]

F_{ij} = force, where "i" subscript is related to the orientation of the face area upon which the force acts; and "j" subscript indicates the direction of the force (Equations 1.16 and 1.17) [Pa]

G = Shear Modulus (Equation 1.14) [Pa]

h = heigh (Equations 1.10 and 1.12) [m]

H = Bulk Modulus (Equation 1.15) [Pa]

L = length of a stressed sample (Equation 1.3) [m]

L_0 = length of an unstressed sample (Equation 1.3) [m]

R = radius (Equation 1.9) [m]

t = time (Equations 1.20, 1.21, 1.22, 1.24) [s]

u = displacement velocity (Equations 1.18, 1.22, 1.23, 1.25, 1.26, 1.27, 1.28, 1.29) [m/s]

x = displacement x-axis (Equations 1.19, 1.21) [m]

y = displacement y-axis (Equations 1.18, 1.22, 1.23, 1.25, 1.26, 1.27, 1.28, 1.29) [m]

References

Augusto, P. and Vitali, A. 2014. Assessing juice quality. In *Juice processing*, pp. 83–136, CRC Press.

Bourne, M. 2002. *Food texture and viscosity: concept and measurement*, Elsevier.

Ibarz, A. and Barbosa-Canovas, G. V. 2014. *Introduction to food process engineering*, Taylor & Francis.

Kramer, A. and Szczesniak, A. S. 1973. *Texture measurement of foods: psychophysical fundamentals; sensory, mechanical, and chemical procedures, and their interrelationships*, Springer Science & Business Media.

Oliveira, M. M. D., Augusto, P. E. D., Cruz, A. G. D. and Cristianini, M. 2014. Effect of dynamic high pressure on milk fermentation kinetics and rheological properties of probiotic fermented milk. *Innovative Food Science & Emerging Technologies*, 26, 67–75.

Penaloza-Giraldo, J. 2019. *Propiedades: Definición de viscosidad*. https://www.youtube.com/channel/UCMlzt73RWcBb5EbKma2zXpw/about

Rao, M. A. 2013. *Rheology of fluid, semisolid, and solid foods: principles and applications*, Springer Science & Business Media.

Fluid-Flow Properties

Steady-State Shear and
Time-Dependent Flow Behaviors

Pedro E. D. Augusto, Meliza L. Rojas and Alberto C. Miano

2.1. Fluid Flow: Steady-State Shear Properties

The rheological properties related to the fluid flow can be understood by the following experiment. Suppose that two parallel plates with a fluid between them (Figure 2.1a). Each plate has an area A in contact with the fluid, being separated by a distance dy. Suppose that there are no border effects (which could be true in a long and wide plate). If the upper plate is moved by a force F, applied on the z direction (Figure 2.1b), after a short transition period, the upper plate will keep moving with velocity v, while the force F still operates on that, even if the lower plate will keep at rest. In this scenario, the fluid layer just below the plate will follow it, moving to the z direction with the same

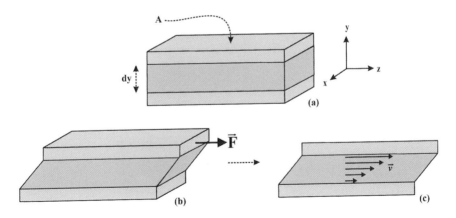

Figure 2.1 *A fluid (in blue) between two parallel plates (in gray) at rest. Each plate has an area A in contact with the fluid, and the distance between the plates is dy (a); a force F is applied to the upper plate, which is moved in the z direction (b); a cross-section in the yz plane, showing the velocity vectors v on the fluid (c).*

DOI: 10.1201/9781003148722-2

v velocity. On the other hand, the fluid layer just above the lower plate will stay at rest. Therefore, the fluid between these two layers will move at the z direction with a velocity profile (Figure 2.1c), as each fluid layer will drag immediately below one.

We can thus define two important properties of fluid flow with this procedure. The first is the shear stress (σ, Equation 2.1), which relates the force F with the applied area A, tangent to the F direction (in the experiment, the area A is on the xy plane, while the force F is on the z direction—Figure 2.1). Then, we can define the shear rate ($\dot{\gamma}$, Equation 2.2), i.e., the fluid velocity gradient across the plates. Some typical shear rate ($\dot{\gamma}$) values in food processing are shown in Figure 2.2.

$$\sigma = \frac{F}{A} \tag{2.1}$$

$$\dot{\gamma} = \frac{dv}{dy} \tag{2.2}$$

It is interesting to notice the shear rate ($\dot{\gamma}$) values vary widely during the different steps and unit operations related to food processing, commercialization and storage—Figure 2.2 presents a difference of 11 orders

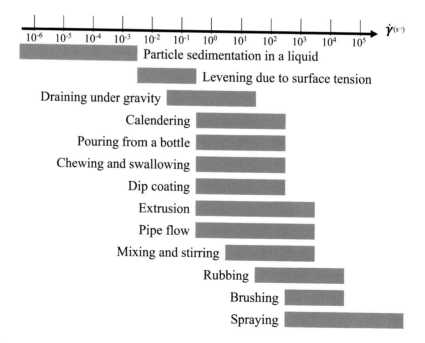

Figure 2.2 *Some typical shear rate ($\dot{\gamma}$) values in food processing.*

Source: Adapted from Steffe (1996)

of magnitude between the extremes. In fact, there is no other process condition or physical quantity that varies so much during processing. Consequently, food products can experience a great range of rheological properties during processing, highlighting the relevance of understanding it in order to improve and develop new processes and products.

Different behaviors would be observed by placing fluids with different structures in the experiment of Figure 2.1.

If a perfect fluid is placed between the plates, the shear rate will be linearly proportional to the applied shear stress (Figure 2.3). Thus, the constant of proportionality is called viscosity (η, Equation 2.3: Newton's Law—Ibarz and Barbosa-Cánovas [2014]; Rao [2010]; Steffe [1996]), and represents the fluid resistance to flow. The fluid with this behavior is called Newtonian, being the behavior of pure substances (in one phase, such as water), perfect mixtures (such as gases and oils) and solutions (such as sugar and salt solutions, or clarified juices). Moreover, some dispersions, when diluted, also present Newtonian behavior, such as milk and diluted juices. However, as the concentration of dispersions is increased and/or their components interact among them, the fluid behavior deviates from Newtonian behavior.

$$\sigma = \eta \cdot \dot{\gamma} \qquad (2.3)$$

In fact, most of the foods show a different behavior under flow. The fluids whose behavior deviates from Newton's Law are called non-Newtonian, being that their structure and composition are responsible for defining their behavior.

When the product is composed of irregular-shaped suspended particles (as the pulp in fruit products, constituted by fruit tissue cells

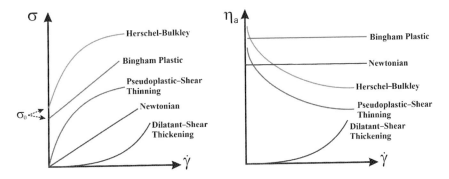

Figure 2.3 *Flow behavior of fluids. Shear stress versus shear rate (left) and the variation of the apparent viscosity as a function of the shear rate (right).*

and their fragments, cell walls and insoluble polymer clusters and chains), three-dimensional polymer clusters and chains (as dispersions of polysaccharides and proteins) and/or non-regular aggregates (as particle clusters, flocks), their spatial distribution is random at rest (Figure 2.4 at left). In this state, the resistance of the fluid to flow is higher. However, when sheared, those suspended structures tend to align to the flow direction, reducing the overall resistance to flow. In this kind of fluid, thus, the resistance to flow decreases in relation to the shear rate increasing, which is called pseudoplastic or shear-thinning fluids.

The pseudoplastic fluids cannot be described by the Newton's Law (Equation 2.3) as its "viscosity" is not a constant property, but a decreasing function of the shear rate. Therefore, the resistance to flow will be characterized by the apparent viscosity (η_a), defined in Equation 2.4. Moreover a new, and more complex model, must be used to describe it.

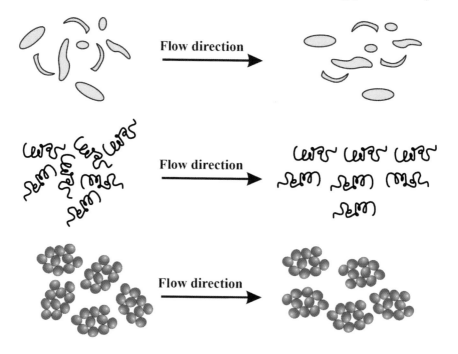

Figure 2.4 *Explanation of pseudoplastic (shear-thinning) behavior, as a result of the particle/molecular alignment to the flow.*

The Ostwald–de Waele Model (Equation 2.5—Ibarz and Barbosa-Cánovas [2014]; Rao [2010]; Steffe [1996]), also called the power law model, describes the fluid flow behavior using two parameters:

- The parameter k, called the consistency coefficient, is related to the fluid consistency and resistance to flow. The k unit is Pa·s^n.
- The parameter n, called the flow behavior index, describes the fluid behavior under shear. n is a dimensionless parameter, being $0 < n < 1$ for pseudoplastic fluids.

It is important to observe that a more complex structure leads to new interactions among its constituents, deviating the fluid behavior from the perfect fluid (Newtonian behavior). Consequently, a more complex mathematical model must be used to describe it. However, the more complex model should also be able to describe the simpler behavior. Therefore, the Ostwald–de Waele Model can describe the behavior of Newtonian fluids when $n = 1$, when Equation 2.5 is reduced to Newton's Law. For pseudoplastic fluids, $0 < n < 1$, which describes a power decreasing behavior of η_a in relation to the shear rate (Figure 2.2).

$$\eta_a = \frac{\sigma}{\dot{\gamma}} \tag{2.4}$$

$$\sigma = k \cdot \dot{\gamma}^n \tag{2.5}$$

The Ostwald–de Waele Model is commonly used to describe the rheological properties of food products, in particular fruit juices and derivatives, explaining well the behaviors described in Figure 2.4. The model parameters also allow an interpretation of the fluid microstructure. When the suspended particles/molecules are perfect spheres, their alignment to the flow is negligible. In this case, the fluid will behave as Newtonian, being $n = 1$. This is the case of milk, a suspension of proteins and emulsion of lipids in a solution of sugar and minerals. Being a dispersion, its behavior is expected to be different from the Newtonian one. However, being in the lipid phase organized as spherical fat globules while the proteins are mainly organized in micelles (with a quasi-spherical shape), the milk behavior is Newtonian. However, by increasing the concentration of milk constituents (such as removing water or producing cream), the concentrated suspension behaves as a pseudoplastic, as a consequence of the deviation of sphericity of its components and interaction among them. This fact can be observed by evaluating the rheological behavior of cream milk with different concentrations and at different temperatures (Figure 2.5). At smaller concentrations and higher temperatures, the distance and interaction of the suspended

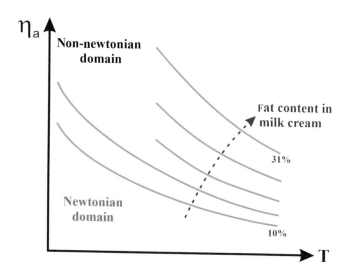

Figure 2.5 *Rheological behavior of milk cream at different fat concentrations: apparent viscosity as a function of temperature, highlighting the conditions of Newtonian and non-Newtonian (pseudoplastic) behaviors.*

Source: Data from Flauzino et al. (2010).

compounds dispersed in the serum are such that the fluid behaves as Newtonian. However, by increasing the fat concentration and reducing the temperature, the alignment effect becomes important, resulting in a pseudoplastic behavior for the milk cream (in this example, it can be seen for temperatures below 40°C and fat concentrations from 20%)—the effect of temperature on the rheological properties is detailed in Chapter 4. Therefore, one can interpret the pseudoplastic behavior, through the parameter n in the Ostwald–de Waele Model, as an indicative of the suspended particles morphology (whose deviation from sphericity results in a deviation from the Newtonian behavior, reducing n).

The opposite behavior is called dilatants or shear thickening, being represented in Figure 2.6. This behavior is characteristic of high-concentrated suspensions with rigid particles. When dilatant fluids are sheared, many collisions of its suspended particles are observed, increasing the overall resistance. In this kind of fluid, thus, the resistance to flow increases in relation to the shear rate increasing (Figure 2.3). There are a few examples of dilatant fluids in special food products, such as highly concentrated starch suspensions, some crystallized honey and suspended sand in water. One can verify some interesting videos on the internet with highly concentrated starch suspensions demonstrating this behavior (search for "non-Newtonian fluids").

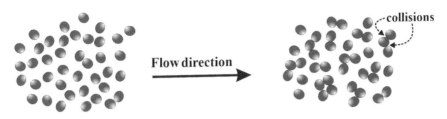

At rest

Flowing (under shear)

-- collisions

Flow direction

Figure 2.6 *Explanation of dilatant (shear-thickening) behavior, as a result of the collisions among rigid particles during flow.*

The Ostwald–de Waele Model (Equation 2.5), also called the power law model, can also be used to describe the dilatant behavior:

- The parameter **k**, called the consistency coefficient, is related to the fluid consistency and resistance to flow. The **k** unit is Pa·s^n.
- The parameter **n**, called the flow behavior index, describes the fluid behavior under shear; **n** is a dimensionless parameter, being **n** > 1 for dilatant fluids.

Furthermore, depending on the interparticle interactions in the suspensions, two other flow behaviors can be seen in food products.

The Peclet number relates the particle transport due to shearing (non-Brownian systems) and diffusion (Brownian systems, Equation 2.6—Fischer et al. [2009]; Rao [2010]). As the particle size or the shear rate is reduced, the *Pe* decreases and the system approximates the Brownian domain. In the Brownian domain, both the electrostatic and van der Waals forces can dictate the interparticle interactions, while only hydrodynamic forces dictate those interactions at higher Peclet numbers (Figure 2.7). Therefore, as the shear rate tends to zero, the fluid is under the Brownian domain, and both the electrostatic and van der Waals forces act in order to maintain the suspended particles stable, forming an internal structure similar to a network. If a stress higher than a limit value is applied to the system, this network collapses, and the fluid starts to flow—which is represented in Figure 2.7.

$$Pe = \frac{\eta_{continuous\_phase} \cdot \overline{r}_{particle}^{3} \cdot \dot{\gamma}}{k_B \cdot T} \tag{2.6}$$

The yield stress (σ_0) is the minimum shear stress required to initiate product flow, being related to the material's internal structure, that

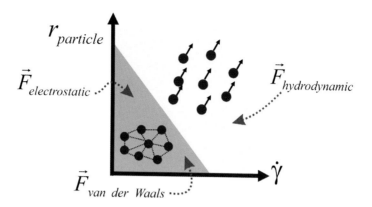

Figure 2.7 *The forces involved in interparticle interactions at different Peclet numbers, as well as the representation of particles states (in small Peclet numbers, at rest, interacting and forming a structure similar to a network; at high Peclet numbers, flowing aligned to the flow).*

must be broken to flow (Genovese and Rao, 2005; Tabilo-Munizaga and Barbosa-Cánovas, 2005). It is a typical characteristic of multiphase materials (Sun and Gunasekaran, 2009), being related to the interactions among suspended particles, droplets and/or molecules. At stress below the yield stress, the material deforms elastically, behaving like an elastic solid, without flowing; above the yield stress, it starts flowing, behaving like a viscous, non-Newtonian liquid. One can interpret the parameter σ_0 as an indicative of the interaction strength among the suspended particles, droplets and/or molecules, as well as the consequent initial structure, similar to a network. Once this barrier is achieved, the fluid can flow with or without the alignment effect in its dispersed phase.

If the fluid presents yield stress (σ_0), and when flowing it does not present the effect of alignment, it is called Bingham fluid or Bingham plastic. In this case, the fluid needs a stress above its yield stress to start flowing, from which it behaves as a Newtonian fluid, with constant apparent viscosity (then called plastic viscosity in the Bingham fluid—η_p; Figure 2.3). There are a few examples of Bingham fluids in special food products, with mayonnaise (and other emulsions) in general described as such. In fact, if one considers the emulsions being formed of spherical lipid droplets dispersed in water (or water solution), the alignment effect of the dispersed phase would be negligible (justifying the flow similar to Newtonian fluids), while the interaction among the emulsifiers would result in the yield stress (σ_0). However, experimental results indicate that emulsions—and mayonnaise in particular—rarely behave as Bingham fluids. At least two points can justify this fact. First, emulsions—and mayonnaise in particular—are made using other components to improve their properties, such as hydrocolloids (polysaccharides and/or proteins) to increase the consistency of the serum phase, increasing its stability to creaming (reducing the velocity of oil droplets separation). Once flowing, those macromolecules are aligned to the flow, as demonstrated in the middle example of Figure 2.4. Second, the dispersed oil droplets are soft and malleable. Consequently, although spherical when at rest, they are progressively deformed to ellipsoids aligned to the flow. Both mechanisms make the fluid present a behavior similar to pseudoplastic fluids when flowing.

When a fluid presents yield stress (σ_0), and when flowing it behaves similarly to pseudoplastic fluids (effect of particle alignment), it is called a Herschel–Bulkley fluid (in general simply called HB fluid). As illustrated in Figure 2.3, it needs a stress above its yield stress to start flowing; it then presents a shear-thinning behavior, with decreasing apparent viscosity in relation to the shear rate. There are many examples of HB foods, such as fruit pulps and juices, which are formed by a dispersion of insoluble components (materials of cellular walls) in a water solution (serum, containing sugars, minerals, proteins and soluble polysaccharides). Other examples are concentrated milk or cream (a suspension of proteins and emulsion of lipids in a solution of sugar and minerals), molten chocolate (dispersion of cocoa seed particles in molten cocoa butter) and—in fact—most of the fluid and semi-solid foods.

It is also important to observe that the amount of dispersed phase in the suspension can change the fluid rheological classification. Fruit juices, for example, can behave as pseudoplastic or Herschel–Bulkley fluids in relation to the amount of pulp and/or to its nature, which defines the interactive forces among them. Milk and cream, as already described in Figure 2.5, pass from Newtonian to pseudoplastic to HB behaviors in relation to the concentration increase. Similar changes are observed by changing the temperature, modulating the mobility and interaction of particles, as detailed in Chapter 4.

If a more complex behavior is observed, as a consequence of a more complex structure, a new and more complex model must be used.

The Bingham Model (Equation 2.7—Ibarz and Barbosa-Cánovas [2014]; Rao [2010]; Steffe [1996]) describes a fluid with yield stress (Figure 2.3) that, when overcome, flow similarly to a Newtonian fluid. Therefore, the Bingham Model adds one parameter for yield stress (σ_0) in the Newtonian Model. Similarly, the Herschel–Bulkley Model (Equation 2.8—Ibarz and Barbosa-Cánovas [2014]; Rao [2010]; Steffe [1996]) describes a fluid with yield stress (Figure 2.3) that, when overcome, flows similarly to a pseudoplastic fluid. Therefore, the Herschel–Bulkley Model adds one parameter for yield stress (σ_0) in the Ostwald–de Waele Model.

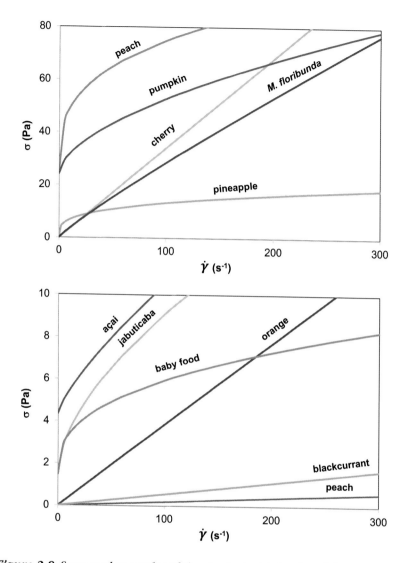

Figure 2.8 *Some real examples of the rheological behavior of fruit juices and derivatives: açai pulp (14% solids, 25°C; data from Tonon et al. [2009]), jabuticaba pulp (14% solids, 25°C; data from Sato and Cunha [2009]); sweet potato baby food (20°C; data from Ahmed and Ramaswamy [2006]); clarified concentrated orange juice (66°Brix, 25°C; data from Ibarz et al. [2009]); clarified concentrated blackcurrant juice (35°Brix, 25°C; data from Ibarz et al. [1992]); clarified concentrated peach juice (12°Brix, 20°C; data from Augusto et al. [2011]); peach puree (25°C, 21°Brix; data from Massa et al. [2010]); pumpkin puree (60°C; data from Dutta et al. [2006]); clarified concentrated cherry juice (66°Brix, 20°C; data from Giner et al. [1996]); Malus floribunda juice (68°Brix, 25°C; data from Cepeda et al. [1999]); and pineapple pulp (11.2°Brix; data from Silva et al. [2010]).*

$$\sigma = \sigma_0 + \eta_p \cdot \dot{\gamma} \qquad (2.7)$$

$$\sigma = \sigma_0 + k \cdot \dot{\gamma}^n \qquad (2.8)$$

It is important to observe again that the more complex model should also be able to describe the simpler behavior. Therefore, the Herschel–Bulkley Model (Equation 2.8) comprises the Newton, Bingham and Ostwald–de Waele (power law) models, describing the fluid flow behavior using three parameters:

- The parameter σ_0, called yield stress, is related to the fluid resistance to initiate the flow. The σ_0 unit is Pa.
- The parameter k, called the consistency coefficient, is related to the fluid consistency, and resistance to flow. The k unit is Pa·s^n.
- The parameter n, called the flow behavior index, described the fluid behavior under shear. n is a dimensionless parameter, being $n = 1$ for Newtonian and Bingham fluids, $0 < n < 1$ for pseudoplastic and HB fluids, and $n > 1$ for dilatant fluids.

Some other mathematical models have been proposed to describe the rheological behavior of fluids, such as the Casson, Mizrahi-Berk, Sisko, Ellis and Vocadlo models, describing the shear stress as a function of the shear rate; and the Cross and Carreau models, describing the apparent viscosity as a function of the shear rate (Steffe, 1996; Rao, 2010). However, most of the fluid can be described using the Herschel–Bulkley Model, as widely used to characterize food products. Consequently, we opted to describe here the models we consider more useful for interpretation of food processes and properties, directing the interested reader to the referenced literature for more examples.

Some examples of fruit juices and derivatives rheological behavior are shown in Figure 2.8, demonstrating with real products the general descriptions of Figure 2.3.

2.2 Time-Dependent Flow Behaviors: Rheopectic and Thixotropic Fluids

The rheological properties of some materials change during time of flowing. The consistency of these materials can increase or decrease along the time under the same conditions (shear rate).

Time dependence is related to the structural change due to shear (Ramos and Ibarz, 1998), i.e., the destruction (Cepeda et al., 1999) and/or re-aggregation of the internal structure during flow. Both behaviors are attributed to the continuous change of the material's internal structure, which can be reversible or irreversible. Consequently, time-dependent rheological characterization is extremely important for understanding the product changes that occur during the process.

The fluid whose behavior shows an increase in the consistency (apparent viscosity) is called rheopectic; the one that shows a decrease in the apparent viscosity is called thixotropic. The factors that contribute to thixotropy also contribute to the pseudoplasticity, and the factors that cause rheopecticity also cause shear thickening (Ibarz and Barbosa-Cánovas, 2014).

Rheopectic behavior is related to the formation or reorganization of the internal structure, which brings with it increasing resistance to flow. It is a rare phenomenon in food products.

On the other hand, thixotropic behavior is very common in dispersions and suspensions, as many food products. Fruit juices, pulps and derivatives are classic examples of thixotropic materials. Figure 2.9 shows an explanation for the thixotropic behavior in foods—for example, fruit juices. In the original product, the internal structure formed by the insoluble pulp dispersed in the serum has a higher resistance to deformation due to the interparticle interactions (and also aggregation), resulting in a higher shear stress. When shearing is carried out, this structure is broken down, as one may notice by the stress decay. Thus, in this kind of fluid, the resistance to flow decreases over the time.

Three models are widely used for describing the thixotropy in foods: Ibarz and Barbosa-Cánovas (2014) and Figoni and Shoemaker (1983)—Figoni–Shoemaker, Equation 2.9; Weltmann (1943)—Weltmann, Equation 2.10; and Hahn et al. (1959)—Hahn–Ree–Eyring, Equation 2.11. The three models describe the shear stress decay over time for an imposed shear rate. It is important to observe that both the Figoni–Shoemaker and Hahn–Ree–Eyring models are essentially similar, with the same mathematical expression (if the exponential function is applied in both sides of the Hahn–Ree–Eyring Model).

$$\sigma = \sigma_e + (\sigma_0 - \sigma_e) \cdot exp(-k_{FS} \cdot t) \tag{2.9}$$

$$\sigma = A_W - B_W \cdot ln\, t \tag{2.10}$$

$$ln(\sigma - \sigma_e) = A_{HRE} - B_{HRE} \cdot t \tag{2.11}$$

From them, we consider the Figoni–Shoemaker Model (Equation 2.10) more intuitive, describing the time-dependent fluid flow behaviors using three parameters through first-order kinetics:

- The parameter σ_0, called initial stress, is related to the shear stress at rest, having Pa as unit.
- The parameter σ_e, called stress at the equilibrium, is related to the shear stress after time enough to reach the equilibrium, having Pa as unit.
- The parameter k_{FS}, is the kinetic parameter, having s^{-1} as unit.

Some examples of fruit juices and derivatives thixotropic behavior are shown in Figure 2.10, demonstrating with real products the general description of Figure 2.9. A typical example of thixotropic food is tomato sauce, affecting some of its applications. For instance, tomato sauce is frequently used to cover pizza dough, which is a challenge in industrial processes. In those processes, the fermented dough is transported through a conveyor belt and injector nozzles spread the sauce over the dough. Every time the injector nozzles open to deposit the sauce, the fluid needs to flow for some time, then return to rest once the dough deposition is discrete. This transient condition increases and decreases the sauce resistance to flow, which can be a challenge for process design. One adopted approach to face it is keeping the sauce under constant flow, recirculating it through the plant, though at the condition of stability, with smaller resistance to flow, as demonstrated in Figure 2.11. This clever approach is a direct application of food rheology knowledge to improve processing.

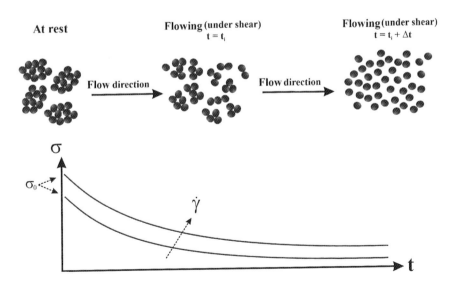

Figure 2.9 *Explanation of thixotropic behavior, as a result of collapse of internal structure.*

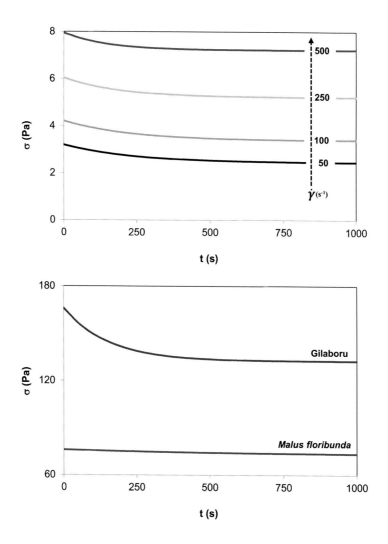

Figure 2.10 Some real examples of the thixotropic behavior of fruit juices and derivatives. Top: tomato juice at 5°Brix, 25°C. (data from Augusto et al. [2012]). Bottom: gilaboru juice at 60°Brix, 20°C (data from Altan et al. [2005]); M. floribunda juice at 72°Brix, 20°C (data from Cepeda et al. [1999]).

Nomenclature

$\dot{\gamma}$ = shear rate [s⁻¹]

η = viscosity (Equation 2.3) [Pa·s]

η_a = apparent viscosity (Equation 2.4) [Pa·s]

η_p = plastic viscosity in the Bingham Model (Equation 2.7) [Pa·s]

σ = shear stress [Pa]

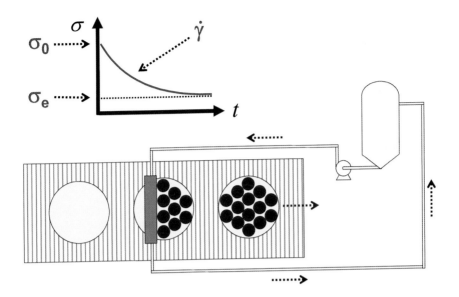

Figure 2.11 *An example of food rheology knowledge application to improve processing: the recirculation of tomato sauce through the plant to keep it at smaller resistance to flow and thus facilitating its spread over pizza dough.*

σ_0 = yield stress, Herschel–Bulkley and Bingham models (Equations 2.7 and 2.8) [Pa]

σ_0 = initial stress in the Figoni–Shoemaker Model (Equation 2.9) [Pa]

σ_e = equilibrium stress in the Figoni–Shoemaker and Hahn–Ree–Eyring models (Equations 2.9, 2.11) [Pa]

A = surface area (Equation 2.1) [m²]

A_{HRE} = structural parameter in the Hahn–Ree–Eyring Model (Equation 2.11) [Pa]

A_W = structural parameter in the Weltmann Model (Equation 2.10) [Pa]

B_{HRE} = kinetic parameter in the Hahn–Ree–Eyring Model (Equation 2.11) [Pa·s⁻¹]

B_W = kinetic parameter in the Weltmann Model (Equation 2.10) [Pa·s⁻¹]

F = applied force (Equation 2.1) [N]

k = consistency coefficient, Herschel–Bulkley and Ostwald–de Waele models (Equations 2.5 and 2.8) [Pa·sⁿ]

k_B = Boltzman constant (Equation 2.6) [= 1.38·10⁻²³ N·m·K⁻¹]

k_{FS} = kinetic parameter in the Figoni–Shoemaker Model (Equation 2.9) [s⁻¹]

n = flow behavior index, Herschel–Bulkley and Ostwald–de Waele models (Equations 2.5 and 2.8) [-]
Pe = Peclet number (Equation 2.6) [-]
$\bar{r}_{particle}$ = mean suspended particle radius (Equation 2.6) [m]
t = time [s]
T = absolute temperature [K]

References

Ahmed, J. and Ramaswamy, H. S. 2006. Viscoelastic and thermal characteristics of vegetable puree-based baby foods. *Journal of Food Process Engineering*, 29, 219–233.

Altan, A., Kus, S. and Kaya, A. 2005. Rheological behaviour and time dependent characterisation of gilaboru juice (Viburnum opulus L.). *Food Science and Technology International*, 11, 129–137.

Augusto, P. E. D., Falguera, V., Cristianini, M. and Ibarz, A. 2011. Influence of fibre addition on the rheological properties of peach juice. *International Journal of Food Science and Technology*, 46, 1086–1092.

Augusto, P. E. D., Falguera, V., Cristianini, M. and Ibarz, A. 2012. Rheological behavior of tomato juice: steady-state shear and time-dependent modeling. *Food and Bioprocess Technology*, 5, 1715–1723.

Cepeda, E., Villarán, M. C. and Ibarz, A. 1999. Rheological properties of cloudy and clarified juice of malus floribunda as a function of concentration and temperature. *Journal of Texture Studies*, 30, 481–491.

Dutta, D., Dutta, A., Raychaudhuri, U. and Chakraborty, R. 2006. Rheological characteristics and thermal degradation kinetics of beta-carotene in pumpkin puree. *Journal of Food Engineering*, 76, 538–546.

Figoni, P. I. and Shoemaker, C. F. 1983. Characterization of time dependent flow properties of mayonnaise under steady shear. *Journal of Texture Studies*, 14, 431–442.

Fischer, P., Pollard, M., Erni, P., Marti, I. and Padar, S. 2009. Rheological approaches to food systems. *Comptes Rendus Physique*, 10, 740–750.

Flauzino, R. D., Gut, J. A. W., Tadini, C. C. and Telis-Romero, J. 2010. Flow properties and tube friction factor of milk cream: influence of temperature and fat content. *Journal of Food Process Engineering*, 33, 820–836.

Genovese, D. B. and Rao, M. A. 2005. Components of vane yield stress of structured food dispersions. *Journal of Food Science*, 70, e498–e504.

Giner, J., Ibarz, A., Garza, S. and Xhian-Quan, S. 1996. Rheology of clarified cherry juices. *Journal of Food Engineering*, 30, 147–154.

Hahn, S. J., Ree, T. and Eyring, H. 1959. Flow mechanism of thixotropic substances. *Industrial & Engineering Chemistry*, 51, 856–857.

Ibarz, A. and Barbosa-Cánovas, G. V. 2014. *Introduction to food process engineering*, CRC Press.

Ibarz, A., Pagán, J. and Miguelsanz, R. 1992. Rheology of clarified fruit juices. II: blackcurrant juices. *Journal of Food Engineering*, 15, 63–73.

Ibarz, R., Falguera, V., Garvín, A., Garza, S., Pagán, J. and Ibarz, A. 2009. Flow behavior of clarified orange juice at low temperatures. *Journal of Texture Studies*, 40, 445–456.

Massa, A., Gonzalez, C., Maestro, A., Labanda, J. and Ibarz, A. 2010. Rheological characterization of peach purees. *Journal of Texture Studies*, 41, 532–548.

Ramos, A. and Ibarz, A. 1998. Thixotropy of orange concentrate and quince puree. *Journal of Texture Studies*, 29, 313–324.

Rao, M. A. 2010. *Rheology of fluid and semisolid foods: principles and applications*, Springer Science & Business Media.

Sato, A. C. K. and Cunha, R. L. 2009. Effect of particle size on rheological properties of jaboticaba pulp. *Journal of Food Engineering*, 91, 566–570.

Silva, V. M., Sato, A. C. K., Barbosa, G., Dacanal, G., Ciro-Velásquez, H. J. and Cunha, R. L. 2010. The effect of homogenisation on the stability of pineapple pulp. *International Journal of Food Science & Technology*, 45, 2127–2133.

Steffe, J. F. 1996. *Rheological methods in food process engineering*, 2nd ed., Freeman Press.

Sun, A. and Gunasekaran, S. 2009. Yield stress in foods: measurements and applications. *International Journal of Food Properties*, 12, 70–101.

Tabilo-Munizaga, G. and Barbosa-Cánovas, G. V. 2005. Rheology for the food industry. *Journal of Food Engineering*, 67, 147–156.

Tonon, R. V., Alexandre, D., Hubinger, M. D. and Cunha, R. L. 2009. Steady and dynamic shear rheological properties of acai pulp (Euterpe oleraceae Mart.). *Journal of Food Engineering*, 92, 425–431.

Weltmann, R. N. 1943. Breakdown of thixotropic structure as function of time. *Journal of Applied Physics*, 14, 343–350.

Viscoelasticity of Fluid, Semi-Solid and Solid Foods

Meliza L. Rojas, Alberto C. Miano and Pedro E. D. Augusto

3.1 Perfect Fluids and Solids and Food Behavior

As previously described, some food products behave as perfect fluids, with a Newtonian rheological behavior, as water, gases, oils, milk, clarified juices, solutions of sugars and other dilute solutions. However, most of the food products show an intermediate rheological behavior between a perfect fluid (i.e., with purely viscous behavior) and a perfect solid (i.e., with purely elastic behavior), and are therefore classified as viscoelastic materials.

Ideal fluids (liquids and gases) respond by deforming continually while the load is applied, and the material does not recover its deformation when the load is removed. The energy involved is dissipated during the deformation and cannot be recovered. This is the called viscous behavior, described by Newton's Law (Equation 3.1).

$$\sigma = \eta \cdot \dot{\gamma} \qquad (3.1)$$

Ideal solids respond by deforming finitely and instantaneously when a load is applied, and immediately recovering that deformation upon removal of the load. The involved energy in this deformation is stored as potential energy, and completely recovered when the material is unloaded. This is called elastic behavior, described by Hooke's Law (Equation 3.2) (Steffe, 1996; Rao, 2013).

$$\sigma = \frac{G}{\gamma} \qquad (3.2)$$

Therefore, viscoelastic materials such as most foods, when deformed, simultaneously show both viscous (fluid) and elastic (solid) components. Figure 3.1 shows the elastic, viscous and viscoelastic behaviors under an applied stress or strain.

A parameter used to characterize or classify substances according to their elastic, viscous or viscoelastic behavior is the Deborah number (De)

DOI: 10.1201/9781003148722-3

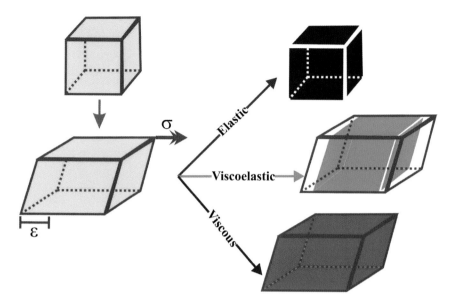

Figure 3.1 *Elastic, viscous and viscoelastic behaviors under an applied stress or strain.*

(Equation 3.3). According to the concept of Deborah number, all substances can flow after waiting the necessary time. In this way, a material can behave like a Hooke solid if it has a very long relaxation time or if it is subjected to a deformation process in a very short time.

$$De = \frac{\tau}{t} \tag{3.3}$$

where t is a characteristic time of the deformation process in which a certain substance is subjected and τ is a characteristic relaxation time of the substance; the relaxation time is infinite for a Hooke solid and zero for a Newton fluid. According to the value of Deborah number, all substances can be classified as one of the following: $De < 1$ (*viscous behavior*), $De > 1$ (*elastic behavior*), $De \approx 1$ (*viscoelastic behavior*).

The microstructure of a product can be correlated with its rheological behavior; the viscoelastic properties are very useful in the design and prediction of the stability of stored samples (Ibarz and Barbosa-Cánovas, 2002). Moreover, viscoelastic products may exhibit some interesting behavior as the Weissenberg and the Barus effects (Ibarz and Barbosa-Cánovas, 2002; Steffe, 1996). Therefore, the study and description of the viscoelastic properties of foods are important for better understanding product behavior and structure during processing, storing and consumption.

3.2 Viscoelastic Properties

Although the steady-state flow procedure is the most valuable way to evaluate the rheological behavior of food products, many phenomena cannot be described by the viscosity function alone, so elastic behavior must be taken into consideration (Steffe, 1996). The steady-state flow procedure has limitations related to slippage and migration of sample constituents (Gunasekaran and Mehmet Ak, 2000), which is particularly true in some food products as fruit juices and dispersions. Moreover, in these experiments, the sample internal structure is broken, limiting the understanding of product behavior in low shear situations, as in particle sedimentation. Therefore, rheological experiments focusing on the product properties at small deformations, whereby the response of a material in this regime is strongly conditioned by its molecular structure, can be used for characterizing those products.

For small deformations, there is a linear relationship between stress and strain called linear viscoelastic range (LVR), which is the range within which the stress is proportional to the applied strain and the theory here described is applied (Gunasekaran and Mehmet Ak, 2000; Steffe, 1996). As the deformation increases, said relationship is no longer linear and the zone of non-LVR is reached, and if the deformation keeps increasing, a point of failure of the structure will occur (Figure 3.2).

Viscoelastic properties of fluid, semi-solid and solid food can be evaluated through three main procedures: the *dynamic oscillatory shear*,

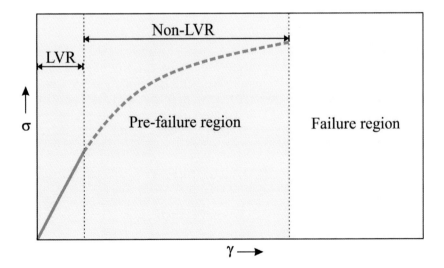

Figure 3.2 *Linear and non-linear relationship between stress and strain in the pre-failure region of structure.*

creep–recovery procedure and *stress–relaxation procedure* (Rao and Steffe, 1992). All these procedures must be carried out in the product LVR, and can be performed by using a rheometer or texture analyzer. In addition, they must be conducted using samples as intact as possible, so pre-tests must be conducted in order to determine the parameters of analysis (see Chapter 5 and Chapter 6).

3.2.1 Viscoelastic Properties: Dynamic Oscillatory Procedure

The dynamic oscillatory procedure consists of applying a small amplitude (typically <5%) (Gunasekaran and Mehmet Ak, 2000) oscillatory movement to the food product. Three variables are involved during the procedure: shear stress (σ), strain (γ) and oscillatory frequency (ω), whereby one of them is kept constant, another is varied and the third is measured. Further, other variables can be varied as the temperature (for melting evaluation, for example) or time (for gelling evaluation, for example).

The main oscillatory procedure used to describe the food product properties is the frequency sweep, whereby the strain (γ) or the shear stress (σ) is kept constant, and its response is evaluated as function of the oscillatory frequency (ω) (these being the three variables within the product LVR). In this experiment, a sinusoidal oscillating movement is applied to the material, and the phase difference between the oscillating stress and strain is measured (Rao, 2013) (Figure 3.3). Thus, the strain over the time varies sinusoidally and can be obtained using Equation 3.4 (where $\gamma_{amplitude}$ is the strain amplitude—Rao [2013]; Steffe [1996]).

$$\gamma(t) = \gamma_{amplitude} \cdot \sin(\omega \cdot t) \tag{3.4}$$

The applied sinusoidal strain input results in a periodic shear stress (σ), which is related to the consequent shear rate ($\dot{\gamma}$) (Equation 3.5; obtained by deriving Equation 3.4—Rao (2013), Steffe (1996), and that is expressed by Equation 3.6 (where $\sigma_{amplitude}$ is the stress amplitude and δ is the phase lag between the stress and strain curves—Steffe (1996).

$$\frac{d\gamma(t)}{dt} = \dot{\gamma}(t) = \gamma_{amplitude} \cdot \cos(\omega \cdot t) \tag{3.5}$$

$$\sigma(t) = \sigma_{amplitude} \cdot \cos(\omega \cdot t + \delta) \tag{3.6}$$

Depending on the phase difference (δ), the viscoelastic properties are determined (Figure 3.3). For ideal solids (i.e., with pure elastic behavior, described by Hooke's Law), the stress and the strain curves are aligned (there is no phase difference: $\delta = 0$). For ideal liquids (i.e., with pure

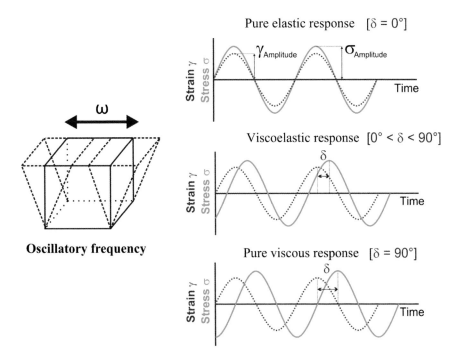

Figure 3.3 *Representation of sinusoidal oscillating movement applied to the material with the elastic, viscous and viscoelastic responses of phase difference between the oscillating stress and strain.*

viscous behavior, described by Newton's Law), the stress and the strain curves are perfectly out of phase: $\pi/2$ ($\delta = 90°$). Therefore, for a product with viscoelastic behavior, the phase lag between the stress and strain curves is between these two values ($0 < \delta < \pi/2$), and the obtained shear stress can be described using Equation 3.7 (Rao, 2013). This equation introduces the parameters G' and G'', whereby the storage modulus (G') is defined by Equation 3.8, and describes the product elastic (solid) behavior indicating the amount of stored energy and released at each oscillatory cycle; while the loss modulus (G'') is defined by Equation 3.9, describing the product viscous (fluid) behavior and indicating the amount of dissipated energy at each oscillatory cycle (Gunasekaran and Mehmet Ak, 2000; Steffe, 1996).

$$\sigma(t) = \gamma_{amplitude} \cdot G' \cdot \sin(\omega \cdot t) + \gamma_{amplitude} \cdot G'' \cdot \cos(\omega \cdot t) \tag{3.7}$$

$$G' = \left(\frac{\sigma_{amplitude}}{\gamma_{amplitude}} \right) \cdot \cos(\delta) \tag{3.8}$$

$$G'' = \left(\frac{\sigma_{amplitude}}{\gamma_{amplitude}}\right) \cdot \sin(\delta) \qquad (3.9)$$

Both storage (G') and loss (G'') moduli are not constant properties, but functions of the strain (γ), oscillatory frequency (ω) and temperature (T). Therefore, the dynamic oscillatory procedures can be carried out considering strain, frequency or temperature sweeps, with different results and interpretations from each one.

In addition, based on G' and G'' values, other viscoelastic functions can be obtained: the loss tangent (tan (δ), Equation 3.10), which describes the ratio of the dissipated and stored energy at each deformation cycle; the complex modulus (G^*, Equation 3.11) and the complex viscosity (η^*, Equation 3.12), which represent the overall resistance of the product to deformation (flow), regardless of whether that deformation is recoverable (elastic, solid) or non-recoverable (viscous, fluid) (Rao, 2013; Steffe, 1996).

$$\tan(\delta) = \frac{G''}{G'} \qquad (3.10)$$

$$G^* = \sqrt{(G')^2 + (G'')^2} \qquad (3.11)$$

$$\eta^* = \frac{G^*}{\omega} \qquad (3.12)$$

Typical frequency sweep results obtained for fruit products such as purees, pulps or juices are shown in Figure 3.4a (Augusto et al., 2012; Augusto et al., 2013a). The magnitude and behavior of the moduli are particular for each food and temperature of analysis; although in general, when $G' > G''$, the product has predominant elastic properties, while when $G'' > G'$, the viscous properties will predominate. The rising tendency of the storage modulus (G') and loss modulus (G'') is commonly modeled as a power function of oscillatory frequency (ω) (Equations 3.13 and 3.14), being useful for describing the viscoelastic behavior of food and dispersions (Rao, 2013). In some cases, it is observed a crossover frequency where the curves of G' and G'' intersect (at this point, the elastic and viscous responses are equal) (Figure 3.4b). The crossover frequency is inversely proportional to the relaxation time (Equation 3.15) (Norton et al., 2010).

$$G' = k' \cdot \omega^{n'} \qquad (3.13)$$

$$G'' = k'' \cdot \omega^{n''} \qquad (3.14)$$

$$\omega_{crossover} = \frac{1}{\tau} \qquad (3.15)$$

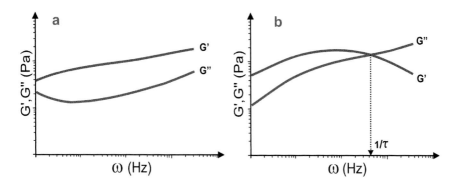

Figure 3.4 *Typical frequency sweep curves at constant temperature for visco-elastic foods.*

Therefore, as both G' and G'' are not constant properties, the product's+ elastic and viscous behavior can be evaluated by the parameters k' and k'', which describe the magnitude of the elastic and viscous comport-ments, and n' and n'', which describe the product elastic and viscous behavior. For example, generally in juices and derivatives, the n' values are lower than the n'' values, which describes that the viscous behavior are more affected for the oscillatory frequency (ω) than the elastic one (Augusto et al., 2013a). Therefore, at lower frequencies, representing the product at rest or very low deformation conditions (as in sedimentation), those products behave as solids, while at higher frequencies, represent-ing the flowing condition, they behave like fluids.

Dynamic rheological tests are generally conducted by small amplitude oscillatory measurements that are non-destructive experiments (Rao, 2014; Dogan and Kokini, 2006). Thus, it is possible to conduct multiple tests on the same sample under different test conditions (Dogan and Kokini, 2006). However, small amplitude oscillatory measurements have the limitation of not being appropriate in practical processing situations due to the low rates and strain at which the test is applied (Dogan and Kokini, 2006; Steffe, 1996). For this characterization, steady-state shear experiments must be conducted (Gunasekaran and Mehmet Ak, 2000). However, as previously stated, the steady-state shear experiments have limitations, such as broken sample structure, limiting the understanding of product behavior in low shear situations, as in particle sedimentation in fruit juices. Dynamic rheological experiments can then be used for characterizing the behavior of those products. Therefore, it is interesting to establish a correlation between the steady-state shear and dynamic oscillatory experiments, using the Cox–Merz rule.

The Cox–Merz rule states that the apparent viscosity ξ_e, at a steady shear rate ($\dot{\gamma}$), is equal to the complex viscosity (η^*) at a specific oscillatory frequency (ω), when $\dot{\gamma} = \omega$ (Equation 3.16—Rao [2013]). When this rule is valid, the rheological food properties can be determined by oscillatory or steady-state shear experiments (Gunasekaran and Mehmet Ak, 2000). It is particularly useful due to the characteristics and limitations in each kind of experiment.

However, the Cox–Merz rule cannot be directly applied in food products whose complex viscosity (η^*) magnitudes are in general higher than the apparent viscosity (η_a) magnitudes, in the whole oscillatory frequency (ω) and shear rate ($\dot{\gamma}$) ranges. Although the Cox–Merz rule has been confirmed experimentally for dispersions and solutions of several polymers (Snijkers and Vlassopoulos, 2014), it is generally necessary in complex systems such as food products to modify the original rule (Gunasekaran and Mehmet Ak, 2000; Rao, 2013). The non-fitting of the Cox–Merz rule for complex dispersions is attributed to structural decay due to the extensive strain applied (Ahmed and Ramaswamy, 2006), presence of high-density entanglements or the development of structure and intermolecular aggregation in solution (Da Silva and Rao, 1992). Therefore, the rheological oscillatory and steady-state shear rheological properties of foods are generally correlated by linear or non-linear modifications of the Cox–Merz rule (Gunasekaran and Mehmet Ak, 2000; Rao, 2013), as those described on Equations 3.17 and 3.18.

$$\eta^*(\omega) = \eta_a(\dot{\gamma})\big|_{\dot{\gamma}=\omega} \tag{3.16}$$

$$\eta^*(\omega) = \alpha\left[\eta_a(\dot{\gamma})\right]\big|_{\dot{\gamma}=\omega} \tag{3.17}$$

$$\eta^*(\omega) = \alpha\left[\eta_a(\dot{\gamma})\right]^{\beta}\big|_{\dot{\gamma}=\omega} \tag{3.18}$$

3.2.2 Viscoelastic Properties: Creep–Recovery Procedure

The creep–recovery procedure is a valuable tool in order to characterize food product behavior, being enabled to correlate it with their structures. By carrying out creep–recovery experiments, it is possible to describe the rheological behavior of the product using mechanical models and constitutive equations, combining Newton's viscosity equation and Hooke's elasticity equations.

In this procedure, a constant stress (σ, in the sample LVR) is applied to the sample and the change in strain (γ) is measured over the time. Then, the stress is instantaneously released and the sample recovery behavior is observed (Steffe, 1996). The typical creep–recovery profile is shown in Figure 3.5. The results are expressed according to the compliance function against time ($J(t)$; Equation 3.19), This method uses

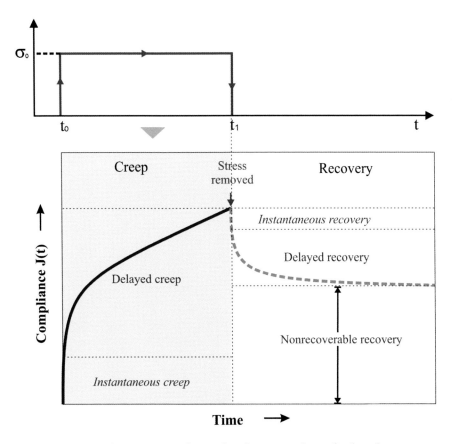

Figure 3.5 *Compliance–time relationship for a viscoelastic food under constant stress.*

fundamental mechanical models to describe the viscoelastic properties (Figure 3.6). A dashpot represents the viscous behavior through Newton's Law (Equation 3.1), whereby the compliance at each instant of time is described by Equation 3.20. On the other hand, a spring, which represents the elastic behavior through Hooke's Law (Equation 3.2), whereby the compliance at each instant of time, is described by Equation 3.21.

$$J(t) = \frac{\gamma(t)}{\sigma_{applied}} \tag{3.19}$$

$$J(t)_{viscous} = \frac{t}{\eta} \tag{3.20}$$

$$J(t)_{elastic} = \frac{1}{G} \tag{3.21}$$

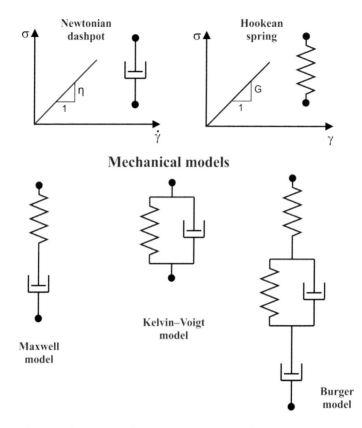

Figure 3.6 *Typical mechanical models, combining dashpots and springs, to describe viscoelastic properties.*

Therefore, the viscoelastic materials can be evaluated and modeled by combining dashpot and springs in series or parallel (Figure 3.6). The most used mechanical models are those of Kelvin–Voigt (a Hookean spring and a Newtonian dashpot placed in parallel), Maxwell (a Hookean spring and a Newtonian dashpot placed in series) and Burgers (a Kelvin–Voigt Model and a Maxwell Model placed in series). Due to product complexity, the Burgers Model best describes food viscoelasticity, as written in Equation 3.22 (Steffe, 1996).

$$J(t)_{Burger} = \frac{1}{G_0} + \frac{1}{G_1} \cdot \left(1 - exp\left(\frac{-G_1 \cdot t}{\eta_1}\right)\right) + \frac{t}{\eta_0} \tag{3.22}$$

It is therefore possible, by using creep–recovery experiments, to isolate and evaluate the viscous and elastic behavior of the material, an important tool to evaluate changes produced by processing or addition of compounds in the preparation of foods such as ice cream (Shama and Sherman, 1966; Sherman, 1966), fruits and derivatives (Augusto et al., 2013b; Nieto et al., 2013), dough and bread (Alba et al., 2020), and emulsions (Samir and Mourad, 2021), among others.

3.2.3 Viscoelastic Properties: Stress–Relaxation Procedure

The stress–relaxation procedure consists of applying a certain strain rate ($\dot{\gamma}$) until a set strain or deformation level (ε) is reached and maintained constant, while the change in stress is measured over time $\sigma(t)$. Therefore, the beginning of the relaxation curve will depend on the level of set strain (γ_0) and the initial stress (σ_0) necessary to reach that strain.

The stress–relaxation curve is composed of both stress–strain relationship and the measured stress over time $\sigma(t)$. In the case of the stress–strain curve, two regions can be differentiated: an independent strain rate region, followed by a dependent region in which an increase in strain rate induces an increase in initial stress. Therefore, the strain level and the strain rate influence the stress–relaxation curves, as represented in Figure 3.7 (Peleg and Calzada, 1976). The *independent strain rate region* occurs at low strain levels where the initial stress (σ_0) and $\sigma(t)$ are proportional to the applied strain. That is, $\sigma(t)$ is a sole function of the strain and independent of the strain rate. In contrast, the *dependent strain rate region* occurs at high levels of set strain (*deformation (ε) > 5%* —Peleg and Calzada, [1976]), whereby the strain rate influences the relaxation behavior.

Therefore, to carry out the stress–relaxation procedure, it is recommended to work at small deformations of the material such that the stress–relaxation curve falls within the *independent strain rate region*. For each food material (depending on its structure), there will be a strain level that describes its viscoelastic behavior in pre-failure stages. Larger failure stress corresponds to stronger structures, but it should be considered that the higher the level of strain applied, the greater the cell damage and water loss which could occur—which would increase the initial decaying rate of the residual stress in the relaxation curves (Tang et al., 1998).

The obtained relaxation curves can be described by different mathematical models, among which the Peleg Model, the Generalized Maxwell Model and Guo–Campanella Model have been successfully applied to describe stress–relaxation behavior in foods.

Peleg (1980) proposed an empirical model that include the normalization of relaxation curves with respect to the initial stress (Equation 3.23).

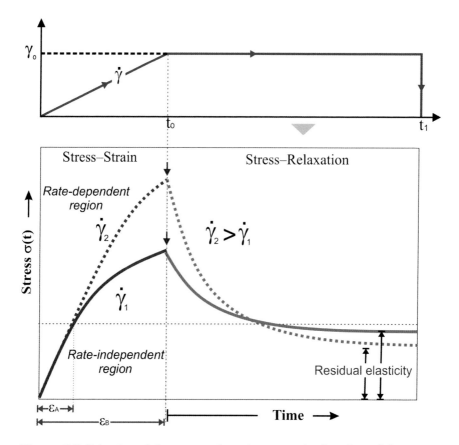

Figure 3.7 *Behavior of the stress–relaxation curve in function of the stress–strain relationship (strain level and rate until reach the set deformation), followed by the measure of stress along time.*

$$\sigma(t)' = \frac{\sigma_0 - \sigma(t)}{\sigma_0} = \frac{t}{k_1 + k_2 t} \tag{3.23}$$

The stress along the time $(\sigma(t))$ is expressed as the sum of the equilibrium stress and a decaying stress (Equation 3.24), whereas the rate of stress decay is expressed according to Equation 3.25 (Tang et al., 1998).

$$\sigma(t) = \sigma_e + \sigma_0 \left(\frac{k_1 / k_2}{k_1 + k_2 t} \right) \tag{3.24}$$

$$\frac{d\sigma(t)}{dt} = -\sigma_0 \frac{k_1}{\left(k_1 + k_2 t \right)^2} \tag{3.25}$$

The initial decay rate of stress ratio is expressed according to Equation (3.26), then $1/k_1$ represents the initial decay rate. The reciprocal of k_2 is the asymptotic level of $\sigma(t)$ when $t \to \infty$. Therefore, for a sufficiently long time, $1/k_2$ will be the representative of the equilibrium conditions (related to equilibrium/residual stress (σ_e) and the residual asymptotic modulus (ξ_e) through Equation (3.27).

$$\frac{1}{\sigma_0}\frac{d\sigma(t)}{dt}\bigg|_{t=0} = -\frac{1}{k_1} \tag{3.26}$$

$$\xi_e = \frac{\sigma_e}{\varepsilon} = \sigma_0\left(1 - \frac{1}{k_2}\right) \tag{3.27}$$

The residual asymptotic modulus (ξ_e) could be used for monitoring induced internal structural modifications, which can have an ambiguous interpretation from the stress–strain relationship alone. For this, different asymptotic moduli can be calculated from performing different relaxation tests at different strain levels. If the asymptotic moduli are plotted against the strain levels, two types of behavior can be observed, as shown in Figure 3.8. The curve (a) shows that ξ_e remains constant until the rupture of the internal structure start, at which point its value begins to decrease. This behavior occurs for almost incompressible (rigid) tissues in a wide range of deformations. On the contrary, curve (b) shows that ξ_e increases with deformation either due to the compressibility of the tissues and/or the development of hydrostatic pressure (Peleg, 1980).

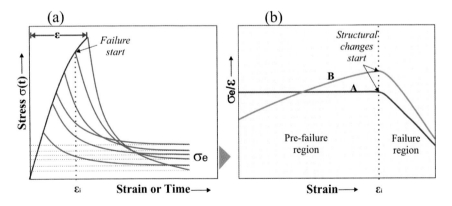

Figure 3.8 *Representation of stress–relaxation curves with different residual stress (σ_e) obtained at different strain levels (a); and of the asymptotic moduli ($\xi_e = \dfrac{\sigma_e}{\varepsilon}$) are plotted against the strain levels (b).*

Source: Based on Peleg (1980)

The Peleg Model (Equation 3.23) has already been used to success-fully describe viscoelasticity of different food products with an excel-lent fitting, and for the same product, changes in values of k_{P1} and k_{P2} parameters could be correlated with modifications in its structure and/or composition (Rojas et al., 2020; Carvalho et al., 2020; Ortiz-Viedma et al., 2018).

On the other hand, the Generalized Maxwell Model (Rao and Steffe, 1992) is a mechanical model that can also be used to describe the relaxation curves. In the Generalized Maxwell Model (GM Model), the Maxwell ele-ments (ME, composed by a Hookean spring and a Newtonian dashpot placed in series, as shown in Figure 3.6) are organized in parallel with an isolated Hookean spring, which contributes to the residual elasticity (ξ_e). However, after a long relaxation time, depending on material type, it can also happen that $\xi_e = 0$ if $\sigma_e \to 0$. This means that the GM Model lacks the isolated spring (Figure 3.9). It is important to note that for the same product and the same fixed deformation (ε), the residual asymptotic modulus (ξ_e) from Equation 3.27 is equal to the residual elasticity (ξ_e) from Equation 3.28, whereby the residual stress $\sigma_e = \xi_e \cdot \varepsilon$, is constant.

Therefore, the GM Model stablishes that stress along the time $\sigma(t)$ dur-ing the stress–relaxation evaluation can be described as a function of the

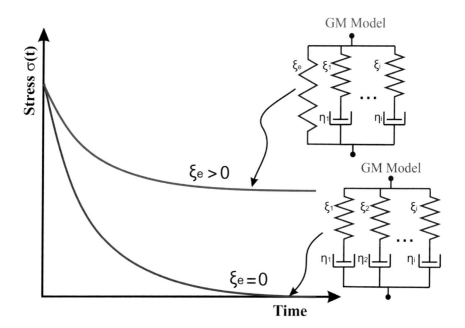

Figure 3.9 *Generalized Maxwell Model to describe stress–relaxation curves with or without residual elasticity.*

constant deformation (ε), the relaxation time (τ_i) and the elastic modulus (ξ_i) (Equation 3.28) (Rao and Steffe, 1992). From the relaxation time and the elastic modulus, the viscous modulus (η_i) was obtained according to Equation 3.29 (Rao and Steffe, 1992).

$$\sigma(t) = \varepsilon.\left(\xi_e + \xi_1.\exp\left(-\frac{t}{\tau_1}\right) + \xi_2.\exp\left(-\frac{t}{\tau_2}\right) + \ldots + \xi_i.\exp\left(-\frac{t}{\tau_i}\right) \right) \qquad (3.28)$$

$$\eta_i = \tau_i.\xi_i \qquad (3.29)$$

The greater the number of ME there are in the model, the better the fit which is obtained; however, this makes it difficult to interpret the parameters. Figure 3.10 shows the stress–relaxation curves obtained from fresh and processed pumpkin (Rojas and Augusto, 2018), which were described by the GM Model (with two Maxwell elements and one elastic element). The GM Model has been decomposed into each of its elements to demonstrate its contribution to describe the behavior of the relaxation curve.

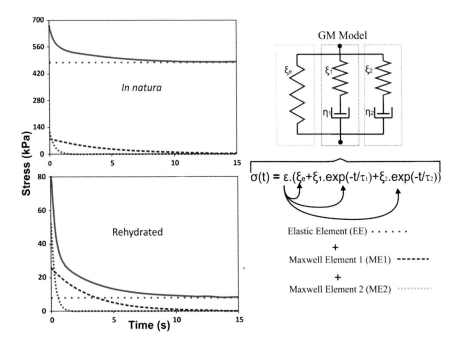

Figure 3.10 *Stress–relaxation curves of fresh (*in-natura*) and rehydrated pumpkin cylinders are shown. Representation of the contribution of the Generalized Maxwell Model elements and their changes in behavior due to structure modification.*

Source: Data from Rojas and Augusto (2018)

The elastic element is related to the elastic constant which depends on the residual stress. On the other hand, the first ME describes a behavior of increasing the initial speed of stress decay in the relaxation curve; however, the second ME describes a much higher stress decay rate than the first ME but a much shorter relaxation time. That is, in the GM Model, the first elements represent a greater contribution to the description of the relaxation curve, while the later elements represent a lower contribution and describe more viscous than elastic behaviors. Therefore, each element and its respective modulus (elastic and viscous) could be related to the structure and composition of a certain material.

In the case of food material, since the parenchymatic microstructure is the main tissue of the plant edible part (Aguilera and Stanley, 1999; Dickinson, 2000), structural elements such as cells, cell walls, intercellular structure and middle lamella could be correlated with the GM Model elements (Rojas and Augusto, 2018), Therefore, the number of Maxwell elements is not specific for all food materials, since it will depend on their original tissues that comprise the structure, composition and changes that occur during processing.

Another model to describe viscoelasticity is the Guo–Campanella Model, which was proposed by Guo and Campanella (2017) to describe the viscoelastic behavior of potato tubers. The model is based on fractional calculus, which is a powerful tool applied to describe complex physical behavior of food materials.

Starting from a general stress–strain relationship based on a differentiation of an arbitrary (fractional) order α with respect to time t, Equation 3.30 was developed to show the relationship between stress α and time t in the fractional calculus approach.

$$\sigma(t) = -I \frac{\varepsilon}{\Gamma(1-\alpha)} \cdot \frac{d}{dt} \int_0^t (t-\tau)^{-\alpha} d(t-\tau) \tag{3.30}$$

After integration of Equation 3.30, the Equation 3.31 was obtained, which describes the relaxation behavior of a viscoelastic food material.

$$\sigma(t) = I \frac{\varepsilon}{\Gamma(1-\alpha)} t^{-\alpha} \tag{3.31}$$

Where, $\sigma(t)$ is the stress over time, I is a parameter that represents the viscoelastic modulus, ε is the strain or deformation (which is constant in the stress–relaxation test), t is the compression time, α is the fractional order and Γ is the gamma function.

Figure 3.11 illustrates the behavior of relaxation curves with a fixed value of I to show the effect of different values of α on $\sigma(t)$. When $\alpha = 0$, Equation 3.31 simplifies to the Hooke's law of elasticity (Equation 3.32)

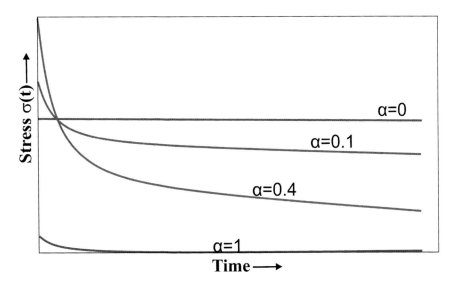

Figure 3.11 *Stress-relation behavior described by Equation 3.31 for different values of fractional order (α).*

Source: Based on Guo and Campanella (2017)

and I represents the elastic modulus. On the other hand, when $\alpha = 1$, Equation 3.31 simplifies to Newton's Law of Viscosity (Equation 3.33) and I represents the viscosity. Finally, when $0 > \alpha < 1$, Equation 3.31 represents the viscoelastic behavior, being I a viscoelastic modulus.

$$\sigma(t) = I \cdot \varepsilon \tag{3.32}$$

$$\sigma(t) = I \cdot \frac{\varepsilon}{t} \tag{3.33}$$

Compared with the number of parameters used in traditional models such as the Generalized Maxwell and Burgers models, the model for stress–relaxation would be simplified by the use of the fractional calculus approach being only two undetermined parameters (α and I) in Equation 3.31. This equation can be applied to describe the relaxation behaviors of *in-natura* (fresh) and processed food viscoelastic materials accurately (Augusto et al., 2018).

3.3 Food Texture

The texture of food is complex to define since it includes different physical characteristics that are physically perceived through the sense of touch when the food is in contact with the fingers or in the mouth, but they can also be perceived by the senses of vision, hearing and kinesthesia (Terefe and Versteeg, 2011). The texture of food as it is perceived and defined by humans (sensorial texture) is difficult to determine by mechanical testing machines. It must be considered that the sensorial texture occurs through a complex mechanism and is not only a mechanical phenomenon, but also other factors occur such as moistening, melting, aroma and flavor release that affect how the texture is perceived (Chen, 2009).

In living organs, such as plant foods, their structure is composed of one or more types of tissues, each characterized by different shape, organization and composition of individual cells. In this type of foods, the texture depends on the structural integrity of tissues and the turgor pressure generated inside the cells by osmosis (Jackman and Stanley, 1995), which changes with any further processing.

The food texture (instrumental texture) can be described by mechanical techniques through different properties which depend on how the structure responds or behaves to an applied stress. If the response does not imply tissue failure (pre-failure region), the behavior can be described by fundamental rheological tests (Wilkinson et al., 2000), through the procedures explained in Section 3.2. On the contrary, if the food is subjected to high deformations, failure of the tissue structure occurs by cell rupture, cell debonding or by a combination of these. Figure 3.12 shows the failure modes of plant tissues under applied stress. The occurrence of one or the other failure mode depends on the middle lamella and the cell wall. If the middle lamella is stronger than the cell wall, cell rupture occurs; in contrast, if the cell wall is stronger, cell debonding occurs (Smith et al., 2003).

The methods and terminology currently used in food science to describe instrumental food texture have been developed empirically since it is difficult to meet the assumptions of phenomenological tests (used in material science) derived from the stress–strain relationship. These assumptions according to Bourne (2002), are: (1) small strains (1–3% maximum); (2) the material is continuous, isotropic (exhibiting the same physical properties in every direction), and homogeneous; and (3) the test piece has uniform and regular shape. However, most textural tests made on foods fail to comply with these assumptions.

Therefore, the terminology used to describe mechanical properties used in material science such as stiffness, strength, toughness and ductility is not widely used to describe the texture of food. Instead, terms like "firmess," "hardness" and "crispiness"—among others—are used. However, these terms are not clearly defined and are only valid within the food literature (Peleg, 2006, 2019), and there are no established methods and conditions to determine them that can be applied to any type of food.

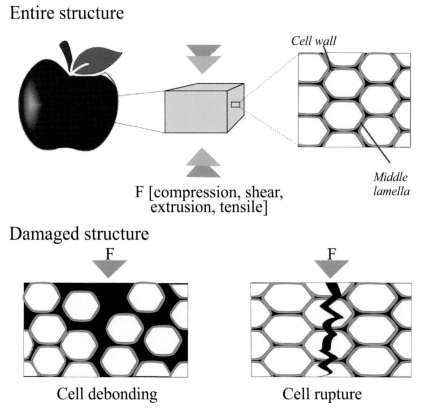

Entire structure

Cell wall

F [compression, shear,
extrusion, tensile]

Middle
lamella

Damaged structure

F

F

Cell debonding

Cell rupture

Figure 3.12 *Failure modes of plant tissues under an applied stress.*

This section will cover destructive techniques involving compression, shear, extrusion and tensile tests since it is considered that destructive tests allow evaluating properties that would have a better correlation with "sensorial texture." This because the texture of a food as it is perceived by humans during chewing occurs in a region of failure or rupture of the food structure (Terefe and Versteeg, 2011; Bourne, 2002). It is important to note that the results obtained from these tests, which are carried out to describe the texture of foods, do not meet the requirements of being independent of the sample size and shape to be called material properties (Peleg, 2006, 2019). Instead of this, it would be better to consider them as texture characteristics of food valid for the type of analyzed food under the actual performed conditions.

3.3.1 Compression and Shear Tests

The most common methods that involves compression and shear forces are compression and puncture tests.

Compression tests are performed by a flat plate or cylindrical plunger (or probe, whose diameter is higher than the sample diameter), depending on the purpose, which can be applied with or without breakdown of structure. Puncture tests, also called penetration tests, are performed by small-diameter cylindrical probes or needle-type probes (with a diameter much smaller than the sample diameter), whereby the puncture force is exerted by a specific point on the food. Both compression and puncture can be performed on the whole product or on a piece of it extracted in a defined geometry (usually cylinders or cubes). It is worth mentioning the importance of selecting an appropriate structure to be evaluated, as well as evaluating how representative is the whole product.

In a typical compression or puncture test, samples are subjected to deformation by compression and shear at constant and relatively low loading rates. Through the compression tests, which result in curves such as the one shown in Figure 3.13, the "firmness" of a fruit or vegetable product can be expressed as the maximum force reached in the force-deformation curves. In some cases, it is possible to identify the rupture point or yield point (the point at which an increase in deformation/strain occurs without an increase in force/stress) which are also used as an indicator of "firmness" (Terefe and Versteeg, 2011; Li et al., 2017).

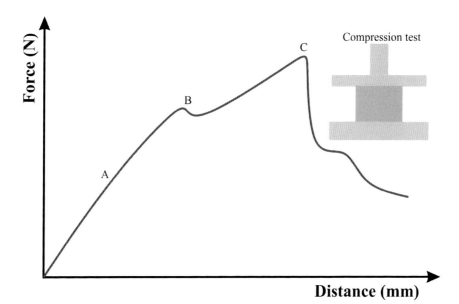

Figure 3.13 *Typical curve obtained with the compression test, showing elastic limit point (A), yield point (B) and massive rupture point (C).*

Source: Based on Li et al. (2017)

In the case of puncture test, penetration of the probe into the food causes irreversible crushing or flowing of the food, and the depth of penetration is usually held constant. Puncture testing instruments may be classed into single-probe instruments such as the Magness–Taylor fruit firmness tester (Abbott, 1999) and the multiple-probe instruments as shown in the pea tester and the Mattson (1946) bean cooker.

By applying the puncture test, different curves were described by Bourne (2002), whereby depending on the structure of the product, different behaviors can be observed (Figure 3.14). The curves A, B and C

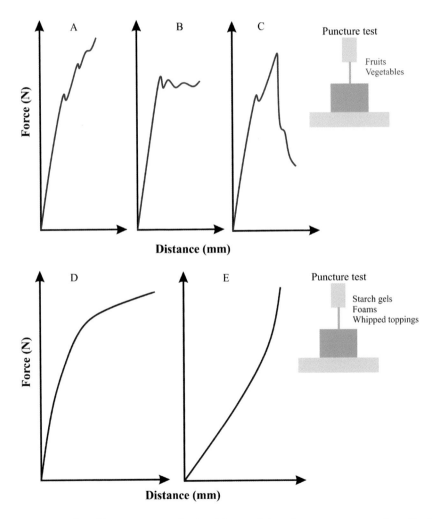

Figure 3.14 *Different curves obtained by applying the puncture test described by Bourne (2002).*

are observed in products such as fruits and vegetables, characterized by a stage of rapid increase in stress or applied force. This stage culminates when the yield point is reached. In berry-type fruits, the yield point would be related to the rupture of the shell. From the yield point, a stage follows that can have three types of behavior: an increase in stress and deformation (A), an increase in deformation without variation in stress (B) and an increase in deformation with a decrease in stress (C). The behavior of this second stage will depend on the internal structure of the analyzed product. Curves C and D are characterized by not clearly showing the starting point of structure failure (yield point), which is observed in products such as starch pastes, gels and products such as foams and whipped toppings.

Through compression testing, it is possible to analyze differences during the ripening stages, storage or differences between varieties or cultivar. For example, Figure 3.15 shows a decrease in the maximum force (at rupture point) with an increase in ripening stage in pseudofruits of *Hovenia dulcis* Thunb. In addition, differences between two apple cultivars (Granny Smith and Golden Delicious) at different times of storage also can be identified.

In addition, there are other devices by which deformation of food samples involves individually or combinations of compression with shear or extrusion forces.

The Kramer shear cell consists of a box and multi-blade shearing device, whereby deformation involves compression, shear and extrusion. The

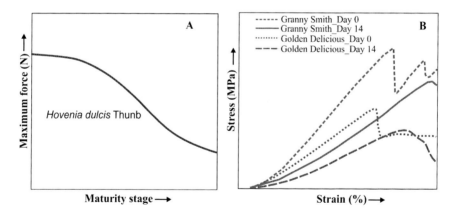

Figure 3.15 *Effects of the ripening stage of pseudofruits of Hovenia dulcis Thunb in the maximum compression force (A) and compression curves of Granny Smith and Golden Delicious apples during storage (B).*

Sources: Data from (A) Maieves et al. (2017); (B) Varela et al. (2007)

sample is loaded in the box and the firmness is expressed as the maximum force per gram or area under the force–deformation curve (Terefe and Versteeg, 2011).

A device that involves compression and extrusion is the dual extrusion cell, in which the test consists of applying force to a food until it flows through an outlet. It can be applied to fluid, semi-solid and solid foods. If the product is solid, initially it is compressed until the structure is disrupted and then extruded. The maximum force required to accomplish extrusion is measured and used as an index of textural quality (Bourne, 2002).

3.3.2 Tensile Test

Tensile testing is not as common as compression for some foods because of the difficulties involved in gripping the sample.

In this method, the sample is often notched at each side through the middle to provide a weakened zone and then locate the failure away from test grips. The grips move apart, stretching the material until it breaks, and the tensile strength is calculated from the maximum force and the cross-sectional area of the sample (Smith et al., 2003; Terefe and Versteeg, 2011).

The drawback of this test it that, in most cases, foods subjected to stress do not fail in a single moment—rather, the fracture begins as a small crack that spreads slowly and may or may not be perpendicular to the plane of applied stress. Therefore, when this type of fracture occurs, it is difficult to objectively interpret the tensile strength. In other cases, tensile tests also can be used to measure the adhesion of a food to a surface. In this case, a disk is placed on a sample (for example, butter), by pressing on it, and then the force required to remove it is recorded (Bourne, 2002).

Nomenclature

α, β = general constant values used in functions that describes rheological parameters [-]
γ = strain (Equation 3.2) [-]
$\dot{\gamma}$ = shear rate (Equations 3.1, 3.16–3.18) [s^{-1}]
δ = phase angle (Equations 3.6, 3.8–3.10) [-]
$tan\delta$ = loss tangent (Equation 3.10) [-]
ξ_i = elastic modulus in Generalized Maxwell Model (Equations 3.28, 3.29) [Pa]
ξ_e = residual elasticity (Equations 3.27, 3.28) [Pa]
ε = deformation level [-]
η = viscosity (Equations 3.1, 3.20) [Pa·s]

η_a = apparent viscosity (Equations 3.16–3.18) [Pa·s]
η_i = viscous modulus in Burgers and Generalized Maxwell models (Equations 3.22 and 3.29) [Pa·s]
η^* = complex viscosity (Equations 3.12, 3.16–3.18) [Pa·s]
σ = shear stress (Equation 3.1) [Pa]
σ_0 = initial stress in Peleg Model (Equations 3.23, 3.24) [Pa]
σ_e = equilibrium stress in Peleg Model (Equations 3.24, 3.27) [Pa]
τ = relaxation time (Equations 3.3, 3.15, 3.28–3.30) [s]
ω = oscillatory frequency (Equations 3.4–3.7, 3.12–3.14, 3.16–3.18) [Hz]
De = Deborah number (Equation 3.3) [-]
G = elastic modulus (Equations 3.2, 3–21) [Pa]
G_0 = instantaneous elastic modulus, associated with the Maxwell spring (Equation 3.22) [Pa]
G_1 = Delayed elastic modulus, associated with the Kelvin–Voigt spring (Equation 3.22) [Pa]
G' = storage modulus (Equations 3.7, 3.8, 3.10, 3.11, 3.13) [Pa]
G'' = loss modulus (Equations 3.7, 3.9, 3.10, 3.11, 3.14) [Pa]
G^* = complex modulus (Equations 3.11, 3.12) [Pa]
I = viscoelastic modulus in Guo–Campanella Model (Equations 3.30–3.33) [Pa, Pa·s]
J = compliance (Equations 3.19–3.22) [Pa^{-1}]
k', k'' = consistency coefficient in power law model of viscoelasticity (Equations 3.13 and 3.14) [Pa·s$^{n'}$, Pa·s$^{n''}$]
k_1, k_2 = parameters of Peleg Model (Equations 3.23–3.27) [s, -]
n', n'' = behavior index in power law model of viscoelasticity properties (Equations 3.13 and 3.14) [-]
t = time [s]

References

Abbott, J. A. 1999. Quality measurement of fruits and vegetables. *Postharvest Biology and Technology*, 15, 207–225.

Aguilera, J. M. and Stanley, D. W. 1999. *Microstructural principles of food processing and engineering*, Springer Science & Business Media.

Ahmed, J. and Ramaswamy, H. S. 2006. Viscoelastic and thermal characteristics of vegetable puree-based baby foods. *Journal of Food Process Engineering*, 29, 219–233.

Alba, K., Rizou, T., Paraskevopoulou, A., Campbell, G. M. and Kontogiorgos, V. 2020. Effects of blackcurrant fibre on dough physical properties and bread quality characteristics. *Food Biophysics*, 15, 313–322.

Augusto, P. E. D., Cristianini, M. and Ibarz, A. 2012. Effect of temperature on dynamic and steady-state shear rheological properties of siriguela (Spondias purpurea L.) pulp. *Journal of Food Engineering*, 108, 283–289.

Augusto, P. E. D., Ibarz, A. and Cristianini, M. 2013a. Effect of high pressure homogenization (HPH) on the rheological properties of tomato juice: viscoelastic properties and the Cox–Merz rule. *Journal of Food Engineering*, 114, 57–63.

Augusto, P. E. D., Ibarz, A. and Cristianini, M. 2013b. Effect of high pressure homogenization (HPH) on the rheological properties of tomato juice: Creep and recovery behaviours. *Food Research International*, 54, 169–176.

Augusto, P. E. D., Miano, A. C. and Rojas, M. L. 2018. Evaluating the Guo-Campanella viscoelastic model. *Journal of Texture Studies*, 49, 121–128.

Bourne, M. 2002. *Food texture and viscosity: concept and measurement*, Elsevier.

Carvalho, G. R., Rojas, M. L., Silveira, I. and Augusto, P. E. D. 2020. Drying accelerators to enhance processing and properties: ethanol, isopropanol, acetone and acetic acid as pre-treatments to convective drying of pumpkin. *Food and Bioprocess Technology*, 13, 1984–1996.

Chen, J. 2009. Food oral processing—a review. *Food Hydrocolloids*, 23, 1–25.

Da Silva, J. L. and Rao, M. 1992. Viscoelastic properties of food hydrocolloid dispersions. *Viscoelastic Properties of Foods*, 285–315.

Dickinson, W. C. 2000. *Integrative plant anatomy*, Harcourt Academic Press.

Dogan, H. and Kokini, J. L. 2006. Rheological properties of foods. In *Handbook of food engineering*, pp. 13–136, CRC Press.

Gunasekaran, S. and Mehmet Ak, M. 2000. Dynamic oscillatory shear testing of foods—selected applications. *Trends in Food Science & Technology*, 11, 115–127.

Guo, W. and Campanella, O. H. 2017. A relaxation model based on the application of fractional calculus for describing the viscoelastic behavior of potato tubers. *Transactions of the ASABE*, 60, 259.

Ibarz, A. and Barbosa-Cánovas, G. V. 2002. *Unit operations in food engineering*, CRC Press.

Jackman, R. L. and Stanley, D. W. 1995. Perspectives in the textural evaluation of plant foods. *Trends in Food Science & Technology*, 6, 187–194.

Li, Z., Miao, F. and Andrews, J. 2017. Mechanical models of compression and impact on fresh fruits. *Comprehensive Reviews in Food Science and Food Safety*, 16, 1296–1312.

Maieves, H. A., Bosmuler Züge, L. C., Scheer, A. D. P., Ribani, R. H., Morales, P. and Sánchez-Mata, M. C. 2017. Physical properties and rheological behavior of pseudofruits of hovenia dulcis thunb in different maturity stages. *Journal of Texture Studies*, 48, 31–38.

Mattson, S. 1946. The cookability of yellow peas. A colloid-chemical and biochemical study. *Acta Agriculturae Suecana*, 2, 185–231.

Nieto, A. B., Vicente, S., Hodara, K., Castro, M. A. and Alzamora, S. M. 2013. Osmotic dehydration of apple: influence of sugar and water activity on tissue structure, rheological properties and water mobility. *Journal of Food Engineering*, 119, 104–114.

Norton, I. T., Spyropoulos, F. and Cox, P. 2010. *Practical food rheology: an interpretive approach*, John Wiley & Sons.

Ortiz-Viedma, J., Rodriguez, A., Vega, C., Osorio, F., Defillipi, B., Ferreira, R. and Saavedra, J. 2018. Textural, flow and viscoelastic properties of Hass avocado (Persea americana Mill.) during ripening under refrigeration conditions. *Journal of Food Engineering*, 219, 62–70.

Peleg, M. 1980. Linearization of relaxation and creep curves of solid biological materials. *Journal of Rheology*, 24, 451–463.

Peleg, M. 2006. On fundamental issues in texture evaluation and texturization—a view. *Food Hydrocolloids*, 20, 405–414.

Peleg, M. 2019. The instrumental texture profile analysis revisited. *Journal of Texture Studies*, 50, 362–368.

Peleg, M. and Calzada, J. 1976. Stress relaxation of deformed fruits and vegetables. *Journal of Food Science*, 41, 1325–1329.

Rao, M. A. 2013. *Rheology of fluid, semisolid, and solid foods: principles and applications*, Springer Science & Business Media.

Rao, M. A. 2014. Rheological properties of fluid foods. In *Engineering properties of food* (Rao, M. A., Rizvi, S. S., Datta, A. K. & Ahmed, J., eds.), 4th ed., CRC Press.

Rao, M. A. and Steffe, J. F. 1992. *Viscoelastic properties of foods*, Elsevier Applied Science.

Rojas, M. L. and Augusto, P. E. D. 2018. Ethanol pre-treatment improves vegetable drying and rehydration: kinetics, mechanisms and impact on viscoelastic properties. *Journal of Food Engineering*, 233, 17–27.

Rojas, M. L., Augusto, P. E. D. and Cárcel, J. A. 2020. Ethanol pre-treatment to ultrasound-assisted convective drying of apple. *Innovative Food Science & Emerging Technologies*, 61, 102328.

Samir, C. and Mourad, L. 2021. Effect of oil-phase volume fraction on rheological properties of pistacia lenticus fruit oil-in-water emulsion intended for healing wounds. *Colloid Journal*, 83, 151–159.

Shama, F. and Sherman, P. 1966. The texture of ice cream 2. Rheological properties of frozen ice cream. *Journal of Food Science*, 31, 699–706.

Sherman, P. 1966. The texture of ice cream 3. Rheological properties of mix and melted ice cream. *Journal of Food Science*, 31, 707–716.

Smith, A. C., Waldron, K. W., Maness, N. and Perkins-Veazie, P. 2003. Vegetable texture: measurement and structural implications. *Postharvest Physiology and Pathology of Vegetables*, 2, 297–329.

Snijkers, F. and Vlassopoulos, D. 2014. Appraisal of the Cox-Merz rule for well-characterized entangled linear and branched polymers. *Rheologica Acta*, 53, 935–946.

Steffe, J. F. 1996. *Rheological methods in food process engineering*, 2nd ed., Freeman Press.

Tang, J., Tung, M. A. and Zeng, Y. 1998. Characterization of gellan gels using stress relaxation. *Journal of Food Engineering*, 38, 279–295.

Terefe, N. S. and Versteeg, C. 2011. Texture and microstructure. *Fruits and Vegetablesh*, 56.

Varela, P., Salvador, A. and Fiszman, S. 2007. Changes in apple tissue with storage time: rheological, textural and microstructural analyses. *Journal of Food Engineering*, 78, 622–629.

Wilkinson, C., Dijksterhuis, G. and Minekus, M. 2000. From food structure to texture. *Trends in Food Science & Technology*, 11, 442–450.

Effect of Product Composition, Structure and Processing on Food Rheological Properties

Meliza L. Rojas, Pedro E. D. Augusto and Alberto C. Miano

4.1 Effect of Product Composition and Structure

4.1.1 Effect of Intrinsic Characteristics: Composition and Structure

The food composition of both the raw material and the industrialized product highly affects its rheological properties. The cell wall and intracellular material composition define the shape and properties of the suspended particles, as well as the product pH and charge. Moreover, the addition of ingredients during processing can change the product's properties. For example, the addition of hydrocolloids in fruit and dairy products is very common in order to minimize sedimentation (pulp, protein, cocoa) and increase sensorial acceptance. Therefore, the intrinsic material characteristics and those changes due to the addition of the ingredients during processing define the final product behavior.

Intrinsic structure and composition of raw materials are influenced by variety, ripening stage at harvest and post-harvest storage time. During fruit ripening, new compounds are generated due to enzymatic activity, such as the increase of simple sugars content and the decrease of polysaccharides (mainly starch and pectin). The breaking of pectin chains from the middle lamella makes cellular debonding easier, then at high ripening stages, the tissues are less hard (Figure 3.15).

The intrinsic characteristics reflect differences in product rheology. Figure 4.1A shows that the apparent viscosity of pineapple juice (4–4.3°Brix) decreases with ripening, while Figure 4.1B shows the apparent viscosity and flow behavior of piñuela juice at two ripening stages and prepared at different soluble solid content. It evidences that rheological behavior depends on the ripening stage of the fruit and the consequent modifications (in quantity and interaction) in soluble and insoluble

DOI: 10.1201/9781003148722-4

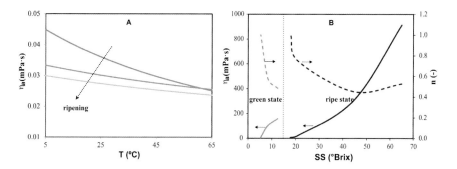

Figure 4.1 *Effect of ripening stage on rheological properties behavior of (A) Josapine pineapple juice (4.3°Brix); and (B) piñuela juice (at $\dot{\gamma} = 100s^{-1}$ and 20°C).*

Sources: Data from (A) Shamsudin et al. (2009); (B) Osorio et al. (2017)

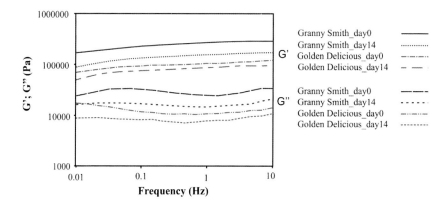

Figure 4.2 *Storage modulus (G') and loss modulus (G'') (at 20°C) of Granny Smith and Golden Delicious apples at different times of storage.*

Source: Data from Varela et al. (2007)

compounds. Figure 4.2 shows the viscoelastic parameters of two apple cultivars, at the initial time and day 14 of storage. It is observed that the level and behavior of the storage and loss moduli depend on the type of cultivar and the day of storage. However, in both cases, with the increase in storage time, the moduli values decrease, which is mainly related to the increase in cell separation due to pectin degradation. This reinforces the differences in fruit products rheology due to their intrinsic characteristics.

4.1.2 Effect of Structure Modifications During Processing

During processing, the tissues and cells of raw materials are disrupted and fragmented, not only increasing the surface area of suspended particles but also changing the properties of the particles and serum. Cell fragmentation exposes and releases wall constituents, such as pectin and proteins, improving the particle–particle and particle–serum interactions and affecting the alignment behavior under flow (Figures 2.4 and 2.7, see Chapter 2). Moreover, the aspect ratio, shape and other characteristics of the suspended particles are changed—not only its mean diameter but also its size distribution (PSD). Once the cell internal constituents are released and put together, chemical and biochemical changes are also induced, due to enzyme activity and also chemical reactions catalyzed by the temperature, modifying the properties of pectin, proteins and other substances. Therefore, the product rheology is also changed, which also modifies how it behaves during processing, storage and consumption.

Products with a broader particle size distribution (PSD), for example, show less consistency than those with a narrow distribution. This is related to the lubricant effect of small particles over larger particles, with a consequent small resistance to flow (Servais et al., 2002). Moreover, particles with smooth surfaces and more regular shapes are easier to flow than those with rough surfaces and irregular shapes.

However, in addition to PSD, the final size of particles will determine the interaction intensity and type of forces which dictate the interparticle interactions and describe the product rheology. The reduction in the suspended particle size can improve interparticle interaction since the particle surface area is greatly increased. The interaction of small particles can be due to van der Waals forces (Genovese et al., 2007; Tsai and Zammouri, 1988) and/or electrostatic forces due to the interaction between the negatively charged pectin and the positively charged proteins (Takada and Nelson, 1983). These interactions are highly affected by pH and temperature (Salminen and Weiss, 2014). Thus, with small particles, the product Peclet number (Equation 2.6) is small and the system approximates to the Brownian domain. In the Brownian domain, both the electrostatic and van der Waals forces can dictate the interparticle interactions, while only hydrodynamic forces dictate those interactions at higher Peclet numbers (Figure 2.7).

Figure 4.3 shows the behavior of the apparent viscosity of apple juice passed through different sieve mesh. It is observed that the smaller the particle size, the higher the apparent viscosity. The increase in apparent viscosity was explained by the lower PSD and the smaller particle size with greater interparticle interaction, although it is important to note that the smaller the mesh opening is, the lower the fraction of pulp (suspended particles) in the product, because more particles are retained on the sieve.

On the other hand, if the number of particles is maintained and only the particle size changes, variations (increase or decrease) in rheological properties can be observed. For example, the effect of particle main size

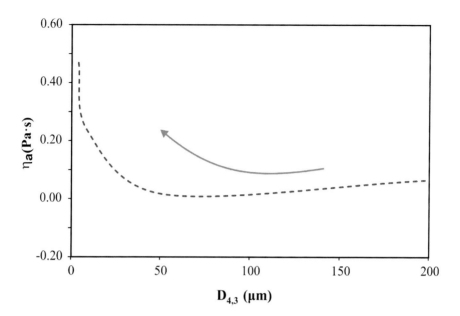

Figure 4.3 *Apparent viscosity (at $\dot{\gamma} = 0.4s^{-1}$ and 25°C) of apple juice in relation to the particle size.*

Source: Data from Zhu et al. (2020)

(at the same particle fraction 21.70% w/w) on the jabuticaba pulp rheological behavior is shown in Figure 4.4, which clearly shows a maximum consistency index (k) at the intermediate value of particle size.

Therefore, even the simplest unit operations—such as pulping, milling, and sieving—can change the final product rheological behavior and must be evaluated in order to obtain the final product with the optimized characteristics.

When the products are frozen, depending on their composition, the cells or the network initially formed breaks due to the ice crystals grown, this change being inversely proportional to the freezing rate. When the cells are disrupted, the internal constituents are released, and cell wall fragmentation also changes the product rheology.

Figure 4.5 shows the behavior of the apparent viscosity of vegetable puree with 5% protein (casein or pea protein) and different hydrocolloids subjected to refrigeration and frozen/thawed. It is observed that, depending on the composition, the apparent viscosity is maintained or decreased, being the product that contains casein and hydrocolloids the most stable after the applied processes. It highlights the potential of using rheology as a tool to study food microstructure and evaluation of technological stability of foods.

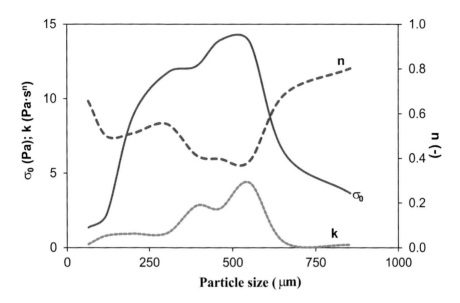

Figure 4.4 *Rheological properties of jaboticaba pulp (13°Brix) in relation to the particle size.*

Source: Data from Sato and Cunha (2009)

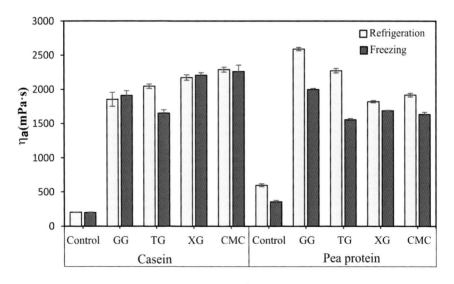

Figure 4.5 *Apparent viscosity (at $\dot{\gamma} = 50 s^{-1}$ and 40°C) of vegetable puree with casein and pea protein (5%) and different hydrocolloids—control, GG (guar gum, 0.95%), TG (tara gum, 0.8%), XG (xanthan gum, 1.2%) and CMC (carboxymethyl cellulose, 1.6%)—under refrigerated and frozen storage.*

Source: Data from Giura et al. (2022)

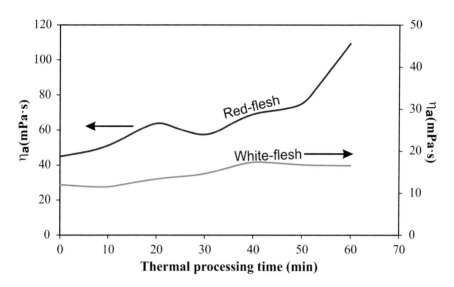

Figure 4.6 *Effect of thermal treatment time (at 70°C) on predicted viscosity of white-flesh and red-flesh dragon fruit purees.*

Source: Data from Liaotrakoon et al. (2013)

The thermal process can also change the product rheology. The consistency can be improved due to the gelation, but it also can be reduced due to pectin damage (as the molecules are heated at low pH). Also, tissues and cells can be disrupted due to thermal or non-thermal effects, as the shear stresses that the product is submitted when pumped through heat exchangers. For example, Figure 4.6 shows an increase in apparent viscosity (measured at 25°C) of dragon fruit puree as the thermal processing time at 70°C increases. This behavior was attributed to polysaccharides that formed a weak gel when time or temperature increases (Liaotrakoon et al., 2013).

This may suggest that dragon fruit has slime-like oligosaccharides which probably form a weak gel with pseudonetwork-like behavior when heating up. There is an implication that a structural change happens while thermal processing is taking place.

The product rheology can also change during storage due to the interactions among different components as the lemon fiber dispersion consistency increases, as shown in Figure 4.7. In fact, age gelation is a well-known phenomenon in products containing soluble fibers, hydrocolloids and—in some cases—proteins. However, an opposite effect can be observed in other types of products because, at a longer time or temperature of the thermal process, the cell walls are disrupted and/or rough particles are smoothed making the product more fluid, as the case observed in Figure 4.1A.

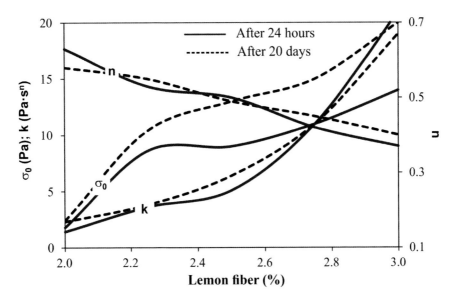

Figure 4.7 *Rheological parameters (at 25°C) of insoluble lemon fiber dispersions at different concentration levels, after 24 hours and after storage (20 days).*

Source: Data from Córdoba et al. (2012)

Therefore, the structure after processing (size, distribution size, shape and morphology) and the coexisting interactions in the medium will determine the rheological behavior of food.

4.1.3 Effect of Composition Modifications During Processing

During processing, the food composition can be modified, whereby acidulants, salts and polysaccharides are added for conservation, sensory or technological purposes. In some cases, water is removed, increasing the product concentration (detailed in Section 4.2.1). Each modification of composition will modify the product properties.

Therefore, the addition of acidulants to foods—a widely used technique to reduce the product pH, microbial growth and enzyme activity—can also affect the final rheology due to the particle electrostatic interactions. Figure 4.8 shows the effect of pH on the rheological parameters of protein solutions. With an increase in pH, the consistency of the protein decreases, while the consistency of the textured protein increases.

Similarly, the salt addition, as well as the addition of ingredients with charge, can highly change the product's rheology. For example, Figure 4.5 shows the effect of the type of salt addition on the apparent viscosity of gum solutions and salt concentration on viscoelastic parameters of

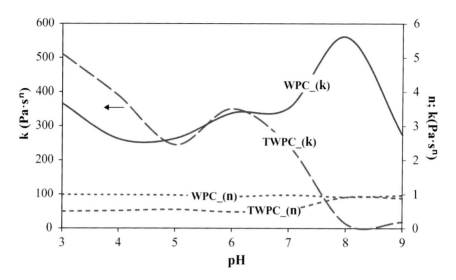

Figure 4.8 *Rheological parameters (at 25°C) of whey protein concentrate 10% w/w (WPC) and texturized WPC (TWPC) with 6% of pre-gelatinized corn starch at pH 3.0–9.0.*

Source: Data from Benoit et al. (2013)

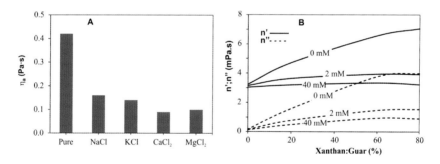

Figure 4.9 *(A) Effect of different salts on apparent viscosity (at $\dot{\gamma} = 46.16s^{-1}$ and 25°C) of 1% w/w Lepidium perfoliatum seed gum. (B) Effect of NaCl concentration on dynamic viscoelastic parameters (at 10 s^{-1} and 20°C) of xanthan:guar blends.*

Source: Data from (A) Koocheki et al. (2013); (B) Khouryieh et al. (2007)

xanthan:guar blends, where it is evident that with the addition of salt and the increase in concentration, the apparent viscosity and the elastic and viscous components decrease.

Similar behavior can be obtained by adding polysaccharides or hydrocolloids. Hydrocolloids are water-soluble gums, extensively used as thickening and gelling agents in food products.

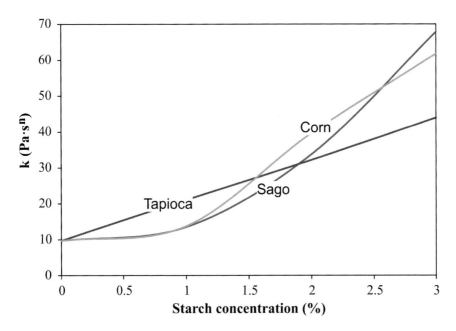

Figure 4.10 *Consistency index (at 37°C) of rice porridges prepared by different thickeners (tapioca, sago, modified corn starch) at concentrations of 1–3% w/w.*

Source: Data from Syahariza and Yong (2017)

Figure 4.10 shows the effect of starch addition (type and concentration) on the rheological properties of rice porridge. The starch addition increases the product consistency, mainly when modified corn and sago starch are used at concentrations greater than 2% w/w.

Hydrocolloids have different thickening capacities to achieve the required rheological properties, and depending on the type of hydrocolloid, different amounts are required (Figure 4.11). Figure 4.12 shows the effect of xanthan gum, guar gum, locust bean gum (LBG), carboxymethylcellulose (CMC) and pectin addition on rheology of fruit juices. Although it is clear the effect of increasing the consistency and pseudoplastic behavior of fruit juices, it is interesting to notice the complex behavior, which is different for each fruit product and cannot be predicted without an entire evaluation. Moreover, due to the possible interactions with other components and at different pH values (as in Figure 4.8) or salt (Figure 4.9), the hydrocolloid addition in foods must be evaluated in each situation.

Therefore, there is still a need for more rheological studies regarding composition effects on food rheology.

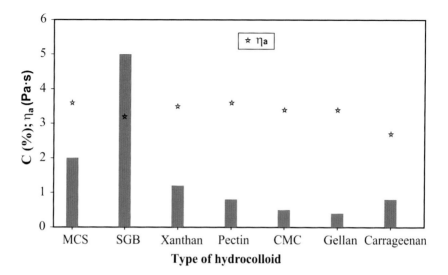

Figure 4.11 *The concentration level of each hydrocolloid added to carrot purees to reach a final apparent viscosity of 3.34±0.31 (at $\dot{\gamma} = 41-50\,s^{-1}$ and 55°C), where: MCS = modified corn starch, SGB = starch–gum blend, CMC = carboxymethyl cellulose.*

Source: Data from Sharma et al. (2017)

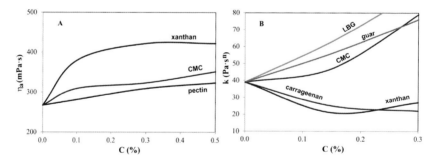

Figure 4.12 *(A) Effect of clouding agents on apparent viscosity (at $\dot{\gamma} = 437.4\,s^{-1}$ and 4°C) of apple juice. (B) Consistency index (at 25°C) of model fruit fillings prepared from waxy corn starch (6.7%) and different concentrations of locust bean gum (LBG), guar, CMC, carrageenan and xanthan gum.*

Sources: Data from (A) Ibrahim et al. (2011); (B) Wei et al. (2001)

4.2 Effect of Concentration and Temperature

4.2.1 Effect of Concentration

The concentration level of a certain compound or food influences its rheological properties exponentially, being usually described using

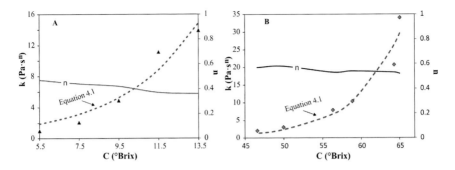

Figure 4.13 *(A) rheological parameters (k and n) (at 20°C) of acerola pulp at different concentrations. (B) rheological parameters (k and n) (at 0°C) of frozen concentrated orange juice at different concentrations.*

Sources: Data from (A) Pereira et al. (2014); (B) Tavares et al. (2007)

Equation 4.1 (Ibarz and Barbosa-Cánovas, 2002; Rao et al., 1999). Figure 4.13 shows some examples of using Equation 4.1 for modeling the influence of solids content on the food rheological properties.

$$A = A_C \cdot exp(B \cdot C) \tag{4.1}$$

Figure 4.13 shows the typical effect of concentration increase on fruit juices flow curves and consistency index (k), which can be well modeled by the exponential function of Equation 4.1.

Figure 4.14 shows the flow curves (shear stress as a function of shear rate) of ready-to-drink peach juice (12.3°Brix) with an addition of 0% (i.e., clarified and depectinized peach juice) to 10% of peach fiber. It is interesting to notice that with fiber addition, not only the consistency but also the product flow behavior is changed. Juices passed from a Newtonian (CF = 0.0%) to a pseudoplastic behavior (CF = 2.5–7.5%), and then to a Herschel–Bulkley's behavior (CF = 10.0%), which is a response to the new interactions among the food constituents.

The influence of the peach fiber addition on the product rheological behavior can be explained by the interactions among the different presented polysaccharides in peach fiber, as well as those among polysaccharides and the natural sugars, acids and water of peach juice. The peach fiber composition is typical of vegetable products, reflecting the vegetable cell wall components, containing soluble and insoluble fractions with cellulose, hemicelluloses, lignin and pectic substances. Thus, the obtained products, like all fruit and vegetable juices with pulp, can be described as insoluble polymer clusters and chains (insoluble or dispersed phase) dispersed in a viscous media composed of

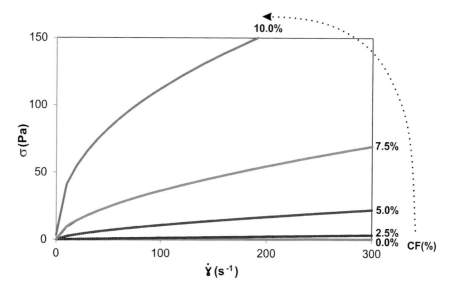

Figure 4.14 *Flow behavior (at 20°C) of clarified peach juice with added fiber concentration.*

Source: Data from Augusto et al. (2011)

soluble polysaccharides, sugars and acids (serum or continuous phase). Therefore, the soluble and insoluble solids concentration and interactions inside each phase and between them can differently impact the product rheology.

The relative viscosity (η_r, Equation 4.2) of dilute solid particles dispersed in a liquid medium is described by the Einstein Equation (Equation 4.3—Genovese et al. [2007]; Metzner [1985]) considering rigid non-interactive particles. Thus, the viscosity of the dispersion is affected by the continuous phase viscosity (the juice serum), the particle intrinsic viscosity ($[\eta]$, which depends on the particle shape) and the particle volume fraction (ϕ). Therefore, the higher product concentration results in higher volume fractions (ϕ) and consequently higher viscosities.

However, the linear relationship described in Einstein's Equation (Equation 4.3) is only valid for dilute dispersions. Moreover, one of the most used models derived for concentrated dispersions is the Krieger–Dougherty Equation (Equation 4.4—Genovese et al., 2007), being more appropriate to describe the food products rheology. According to this equation, the viscosity of the dispersion is also affected by the particle maximum packing fraction of solids (ϕ_m), showing an abrupt increase at $\phi \approx \phi_m$ (Figure 4.15).

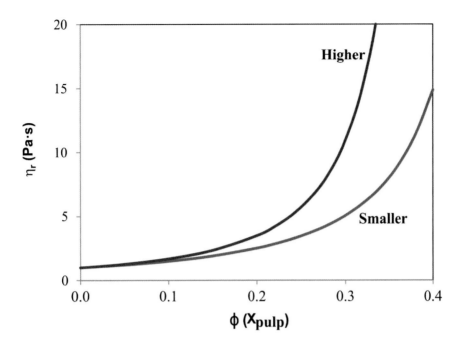

Figure 4.15 *Relative viscosity ($\dot{\gamma} = 20s^{-1}$ and 25°C) of tomato puree in function of pulp weight fraction for two particle sizes.*

Source: Data from Yoo and Rao (1994)

$$\eta_r = \frac{\eta_{dispersion}}{\eta_{continuous\_phase}} \quad (4.2)$$

$$\eta_r = 1 + [\eta] \cdot \phi \quad (4.3)$$

$$\eta_r = \left(1 - \frac{\phi}{\phi_m}\right)^{-[\eta] \cdot \phi_m} \quad (4.4)$$

However, it is important to highlight that the suspended particles in food products are not rigid and non-interactive particles, which can deviate the actual rheological behavior from those stated in Equations 4.2–4.4. Nevertheless, those equations are very useful to understand the role of solids concentration on the rheological properties.

On the other hand, although the increase in product concentration exponentially increases the product consistency index (k) and yield stress (σ_0) due to the interactions among the product constituents, the flow behavior index (n) shows a particular trend. It is generally assumed to be relatively constant with concentration (Rao et al., 1999), which is true just in a narrow-range evaluation (Figure 4.13B). However, it is expected that the increase in product concentration results in changes at rheological fluid behavior (i.e., in the flow behavior index—*n*), whose magnitude may differ from the Newtonian flow. Therefore, the *n* tends to be reduced due to the product concentration, increasing the fruit product's shear-thinning behavior.

However, Figure 4.16 shows a more complex trend of the flow behavior index (*n*) when peach fiber is added to the clarified peach juice. In contrast, with the continuous decrease of an exponential function, it shows a sigmoidal behavior. When the fiber amount is relatively low, *n* shows quasi-constant values, close to the initial Newtonian behavior (i.e., *n* = 1), followed by a great decrease in intermediate fiber concentrations. Then, in a relatively high amount of fiber, the flow behavior index tends to stay back constant. This behavior can be described by a power sigmoidal function (Equation 4.5).

$$n = \alpha + \frac{1-\alpha}{1+\left(\beta \cdot C\right)^{\lambda}} \tag{4.5}$$

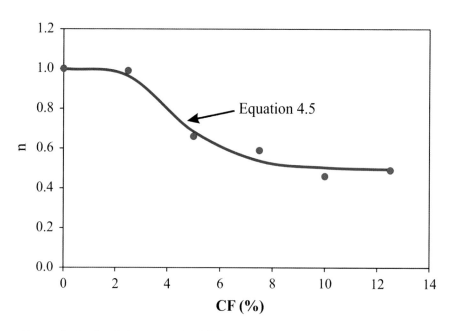

Figure 4.16 *Flow behavior index (n) (at 20°C) in relation to fiber concentration described by a power sigmoidal function (Equation 4.5).*

Source: Data from Augusto et al. (2011)

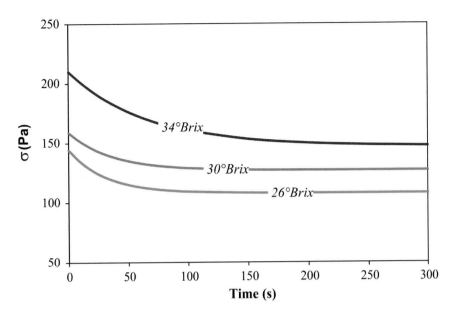

Figure 4.17 *Thixotropic behavior (at 20°C) of peach pulp at different concentrations.*

Source: Data from Lozano and Ibarz (1994)

Regarding product thixotropy, it was reported that an increase in concentration decreases the kinetic parameter and increases the values of initial and equilibrium stresses from Figoni–Shoemaker Model (Equation 2.10) (Figure 4.17), as was reported by Massa et al. (2010) and Lozano and Ibarz (1994). Although the expected behavior is similar to that observed in the steady-state shear properties, there is still a need for a better understanding of the effect of product concentration on time-dependent and viscoelastic properties of food products.

Therefore, the most appropriate approach consists in evaluating each of the fundamental parameters (σ_0, k, n) separately; also, more studies should be carried out in order to better describe the viscoelastic and time-dependent properties.

4.2.2 Effect of Temperature

Temperature is an indirect measure of the product internal energy, greatly affecting the rheological properties. Higher temperatures represent a higher level of internal energy, with a higher distance between molecules, which facilitates molecular movement and vibration, leading to less consistency.

This decrease in consistency follows, in general, an exponential function, being well described by the Arrhenius Equation (Equation 4.6, whereby each rheological parameter A is modeled by a pre-exponential factor (A_0), and the activation energy (E_a), R is the constant of the ideal gases and T is the absolute temperature (Rao, 2014). This mathematical model is successfully used in order to describe food products' viscosity (η), apparent viscosity (η_a), yield stress (σ_0), consistency coefficient (k) and the consistency coefficients related to the viscoelastic properties k' and k''. Figures 4.18–4.20 show some examples of the behavior of rheological properties with temperature.

$$A = A_T \cdot exp\left(\frac{Ea}{R \cdot T}\right) \tag{4.6}$$

Figure 4.18 shows the influence of the temperature on the behavior of the flow curve and rheological parameters (k and η_a) of acerola pulp (13.5°Brix). Similarly, Figure 4.19 shows the variation of the rheological parameters with the temperature in pumpkin puree. The decrease in apparent viscosity and consistency values is observed as the temperature increases, which could be described by the Arrhenius Equation.

The effect of temperature on the peach juice (clarified + 10% of peach fiber) viscoelastic properties is shown in Figure 4.20. The parameters k', k'', n' and n'' could be well described by the Arrhenius Equation.

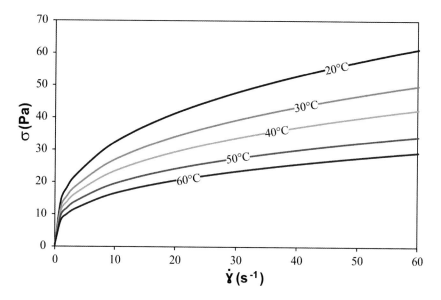

Figure 4.18 *Effect of temperature on flow behavior of acerola pulp (13.5°Brix).*

Source: Data from Pereira et al. (2014)

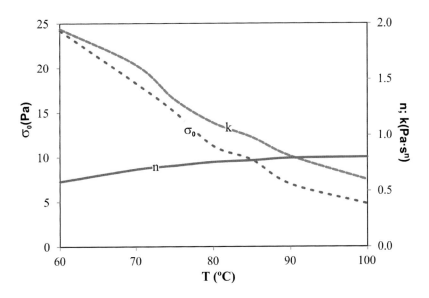

Figure 4.19 *Pumpkin puree rheological parameters as a function of temperature.*

Source: Data from Dutta et al. (2006)

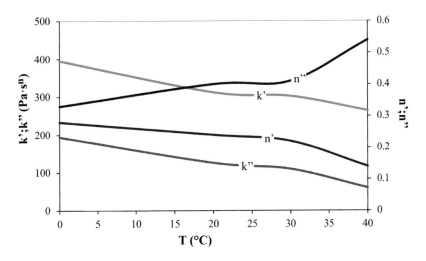

Figure 4.20 *Behavior of viscoelastic parameters* (k', k", n' and n") *with temperature for clarified peach juice with 10% of peach fiber.*

Source: Data from (Augusto et al., 2011)

Moreover, it is interesting to observe that while the values of *n'* decrease with temperature, *n"* shows the opposite behavior. These tendencies clearly show that the product's viscous behavior becomes more important when heated.

Figure 4.21 and Figure 4.22 show the rheological behavior of concentrated citrus juices at low temperatures. Citrus juices are concentrated until 60–66°Brix and then "frozen" at temperatures close to −18°C to be stored and distributed. The clarified orange juice shows a Newtonian behavior, whose viscosity (η) follows the Arrhenius Equation (Figure 4.21). In contrast, the orange juices containing pulp show pseudoplastic behavior, whose consistency coefficient (k), follows the Arrhenius Equation (Figure 4.21 and 4.22), but whose flow behavior index (n) showed a quasi-constant trend (Figure 4.22).

The flow behavior index (n) is generally assumed to be relatively constant with temperature (Rao et al., 1999). However, as the temperature is increased, the molecular mobility results in the product flow behavior index (n) close to the unit (Figure 4.19), as the molecules and particles can be easily aligned (Figure 2.4) and are less susceptible to collision (Figure 2.5). In fact, it is important to observe that the property behavior in relation to temperature is a function not only of the product itself but also of the studied temperature range (i.e., the property value can vary just slightly at the studied temperature range, which is different to be constant in relation to the temperature). When this property is not assumed to be constant, its increasing is modeled using the Arrhenius Equation or even a linear function.

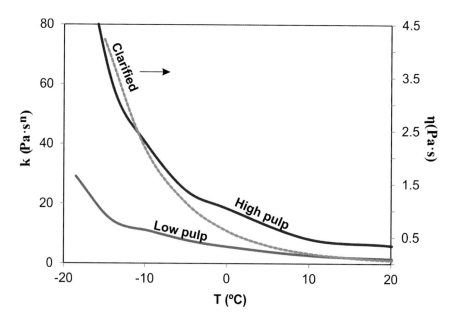

Figure 4.21 *Effect of lower temperatures on consistency index of high pulp and low pulp orange juice (65°Brix) (data from Rao et al. [1984]) and clarified orange juice (66°Brix) (data from Ibarz et al. [2009]).*

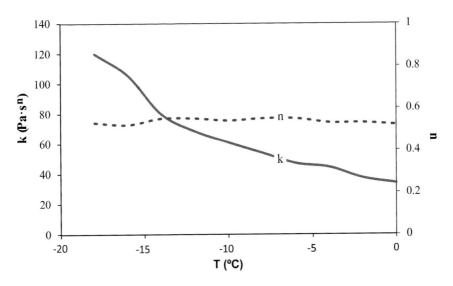

Figure 4.22 *Effect of sub-zero temperatures on consistency and flow behavior index of concentrated orange juice (65°Brix).*

Source: Data from Tavares et al. (2007)

However, other different behaviors from the exponential decay of the Arrhenius Equation can be observed. First, the observed profile only represents the product behavior within the evaluated temperature range. Second, due to the reactions and changes in the product structure due to the consequent thermal process, such as gelling (starch, other biopolymers), vaporization (sample partial drying) and melting (lipids, crystals), among others. Figure 4.23 shows some examples of fruit products rheological properties whose behavior in relation to the temperature does not follow the Arrhenius Equation.

Figure 4.23A shows the results for siriguela (*S. purpurea*) pulp, a small native fruit of Central America. The yield stress shows a falling sigmoidal trend in relation to temperature, in contrast with the continuous decrease of the exponential function (Arrhenius Equation). Until 40°C, the yield stress shows quasi-constant values, followed by a great decrease in temperatures between 40°C and 60°C. Then, it tends to stay back in a constant value. This behavior can be described by a power sigmoidal function (Equation 4.7) and shows that an important change in siriguela pulp appears between 40°C and 60°C (for example, fat melting), as the yield stress value changes are more important in this temperature range. The viscoelastic analysis corroborates this observation, with the parameters *k'* and *k''* showing the same behavior. Therefore, structural changes due to the thermal processing can be evaluated by using rheology.

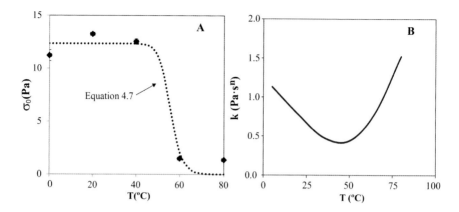

Figure 4.23 *(A) Yield stress behavior of siriguela pulp (10.9°Brix) in relation to temperature. (B) Consistency index behavior of sweet potato baby food (11.2°Brix) in relation to temperature.*

Sources: Data from (A) Augusto et al. (2012a); (B) Ahmed and Ramaswamy (2006)

$$\sigma_0 = \frac{\alpha}{1+(\beta \cdot T)^\lambda} \tag{4.7}$$

Figure 4.23B shows the interesting effect of temperature on the sweet potato baby food consistency coefficient (k). The k value shows a decreasing behavior until 50°C and shows the opposite trend up to this temperature. This is the typical behavior of starch-rich products, such as the sweet potato puree, being related to the gelatinization and consequent consistency increase. Moreover, it is similar to that observed in products with other polysaccharides and proteins with gelation properties. However, it is the opposite behavior of those products whereby melting processing is related to heating. Therefore, it reinforces the need for a better understanding of each food product properties, which can change during processing.

Finally, the effect of temperature and concentration can be modeled by combining the Arrhenius Equation (Equation 4.6) with the exponential function of the concentration (Equation 4.1). The derived function is shown in Equation 4.8, which is extensively used to model the combined effect of concentration and temperature on rheological behavior of fruit products (Ibarz and Barbosa-Cánovas, 2002; Rao et al., 1999).

$$A = A_{TC} \cdot exp\left(B \cdot C + \frac{Ea}{R \cdot T} \right) \tag{4.8}$$

Therefore, because thermal processing, freezing, and refrigeration are unit operations widely used to process and preserve foods, the need is highlighted to conduct more studies regarding the food products' rheological properties at higher and lower temperatures. In fact, due to the difficulty in conducting the experiments at higher temperatures (mainly due to water vaporization), there are only a few works in the literature with rheological evaluation at temperatures above 100°C. Rao et al. (1999) needed to place the rheometer inside a retort, while Ros-Polski et al. (2014) developed a pressurized capillary rheometer in order to conduct their experiments. Therefore, there is still a need for more studies to better understand the effects of temperature on the rheological properties of foods.

4.3 Effect of Emerging Technologies

The effect of the main emerging technologies that cause modification in the structure and interaction of the particles during processing affecting the rheological properties must be known. Then, the effects of the application of ultrasound (US), pulsed electric fields (PEF), high hydrostatic pressure (HHP) and high-pressure homogenization (HPH) are presented.

During the application of these technologies, in addition to the intrinsic characteristics of the food, the food concentration, the temperature and the time of processing, other variables must be controlled or evaluated for each technology since they may affect the rheological properties (Table 4.1).

Food rheology can show complex behavior during ultrasound processing. Rheological properties can be changed permanently or temporarily, either increasing or decreasing, depending on the applied ultrasonic

Table 4.1 Variables That Can Affect Rheological Properties during Processing with Emerging Technologies

US	PEF	HPH	HHP
• Wave frequency • Wave amplitude • Acoustic power	• Electric field strength • Number of pulses • Amplitude of electric pulses • Pulse width • Pulse repetition frequency	• Flow rate • Gap size • Pressurization • Number of cycles	• Compression and decompression rates • Static pressure
	Food characteristics (structure, composition, concentration, pH)		
	Processing time		
	Temperature		

energy (Soria and Villamiel, 2010). For example, the fruit and vegetable fluids processed with ultrasound in the steady-state rheological properties analysis show a non-Newtonian behavior related to the strong interactions among pulp particles. A pseudoplastic flow behavior with a tendency to increase the apparent viscosity was reported in diluted avocado puree after US processing (Bi et al., 2015). Herschel–Bulkley behavior was reported in peach juice (Rojas et al., 2016) and kiwi juice (Wang et al., 2020), whereby variations in yield stress, apparent viscosity, consistency index and flow behavior index in function of processing time were observed.

In fact, what determines the rheological behavior throughout the ultrasound processing time are the modifications in the structure and the interaction of the particles and molecules existing in the continuous and dispersed phase of the product. Figure 4.24 shows the ultrasound effect on the particle size reduction and the changes in the flow behavior of kiwi juice. In this product, it was reported that apparent viscosity initially increased, then decreased and finally increased again with the processing time, including changes in the yield stress and flow behavior (Wang et al., 2020).

The complex rheological behavior as a function of the ultrasound processing time in peach juice was described in detail by Rojas et al. (2016). Variations in the consistency and yield stress were observed, related to modifications in the microstructure which occurs in different stages as the US processing time increases. Structural modifications in the dispersed phase included swelling of cells and movement of intracellular compounds, as well as partial rupture of the cell walls with release of

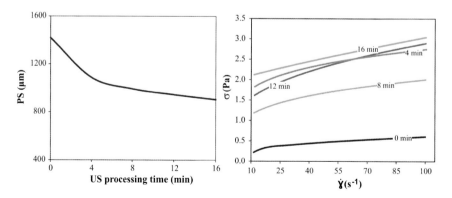

Figure 4.24 *Ultrasound processing (400 W, 20 kHz) time effect on the particle size reduction and the consequent changes in flow behavior (at 25°C) of kiwi juice.*

Source: Data from Wang et al. (2020)

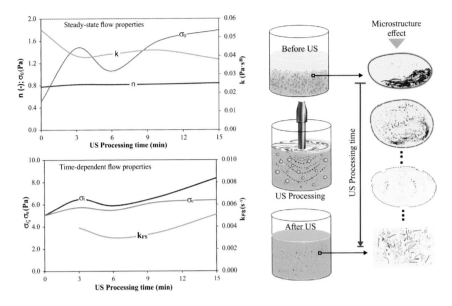

Figure 4.25 *Ultrasound processing (6.67 W/mL, 20 kHz, 22±3°C) time effect on the steady-state and time-dependent rheological properties (at 25°C) of peach juice.*

Source: Data from Rojas et al. (2016)

compounds until complete rupture of the cells (Figure 4.25). In addition, the continuous phase (serum) was also modified with US processing, decreasing its viscosity by 23%.

On the other hand, the time-dependent rheological properties evaluation indicated that although the control did not show thixotropy, the juice becomes more thixotropic after US application by increasing the initial shear stress, the equilibrium stress and the kinetic parameter of the Figoni–Shoemaker Model (k_{FS}, Equation 2.9) (Rojas et al., 2016).

Regarding the application of PEF in food, the main effect is the inactivation of micro-organisms and extraction of compounds; it also causes reversible or irreversible modifications of the cell walls, causing the release of intracellular compounds and even causing cell disruption and breakdown of weak linkages.

The rheological properties change as a function of the time, number of pulses and the electric field strength. In addition, due to the particular configuration of tissues, cells and components, each product behaves different with PEF processing. For example, Figure 4.26 shows the apparent viscosity increase and variations in the consistency and yield stress as the field strength increased (from 0–28 kV/cm). These variations

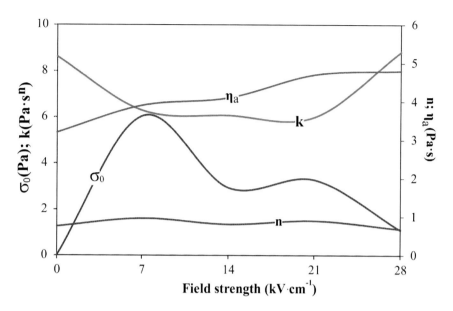

Figure 4.26 *Electric field strength (for 200μs at < 35°C) effect on rheological parameters (η$_a$ at $\dot{\gamma} = 90s^{-1}$) of almond "milk."*

Source: Data from Manzoor et al. (2019)

reflect the changes in the PSD and intermolecular interactions between adjacent denatured molecules of almond milk.

HHP processing modifies the rheological properties of the fluid (processed packed), semi-solid or solid foods since it causes disruption and denaturation of cell membranes and proteins modifying molecular interaction. Figure 4.27 shows the increase and decrease of aloe vera consistency and yield stress as the pressure level increase. Modification in enzyme activity by HHP also affects rheological properties; for example, Hsu (2008) reported that the consistency of tomato juice decrease at 100 MPa and then linearly increases at higher pressure.

HPH applied to fluid foods influenced the rheological properties due to the intense shear stress, sudden pressure drop, cavitation, turbulence, impingement and temperature increase, which affect the particles, molecules and enzymes. Figure 4.28 shows the flow behavior of tomato juices processed by HPH, increasing the resistance to flow with an increase in applied pressure (Augusto et al., 2012c). This makes sense since it was observed a large amount of small particles, composed of cell walls and internal constituents suspended in the juice serum, also resulting in small particles aggregating, forming a network, which must be broken to start to flow. The opposite behavior was described in cashew apple juice.

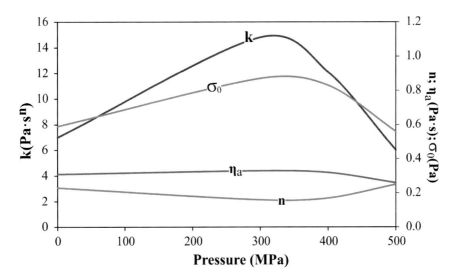

Figure 4.27 *Effect HHP processing on rheological parameters of packed aloe vera suspension processed at different pressure levels (at 15°C for 1 minute).*

Source: Data from Opazo-Navarrete et al. (2012)

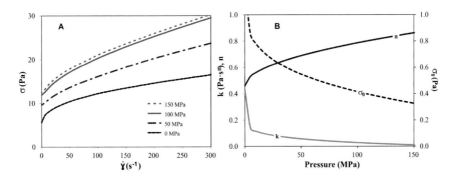

Figure 4.28 *High-pressure homogenization (HPH) processing effect on (A) flow behavior of tomato juice (4.5°Brix, < 40°C); (B) Rheological parameters of cashew apple juice (10°Brix, inlet T = 25°C).*

Sources: Data from (A) Augusto et al. (2012b); (B) Leite et al. (2015)

When the cashew apple juice is processed by HPH, the cell disruption results in particles with a less rough shape and a considerable amount of very small particles whose resistance to flow is smaller than the original product (Leite et al., 2015). It highlights once more the need to evaluate each food product independently, and that it is not possible to predict general behavior.

Nomenclature

α, β, λ = general constant values used in the power sigmoidal functions (Equations 4.5 and 4.7) [-]

$\dot{\gamma}$ = shear rate (Figures 4.14, 4.18, 4.24, 4.28A) [s^{-1}]

η = viscosity [Pa·s]

η_a = apparent viscosity [Pa·s]

η_r = relative viscosity (Equations 4.2–4.4) [Pa·s]

$[\eta]$ = intrinsic viscosity (Equations 4.3, 4.4) [Pa·s]

σ = shear stress (Figures 4.14, 4.17, 4.18, 4.24, 4.28A) [Pa]

σ_0 = yield stress from Herschel–Bulkley Model (Figures 4.4, 4.7, 4.19, 4.25, 4.26, 4.27, 4.28A) [Pa]

σ_i = initial stress from the Figoni–Shoemaker Model (Figure 4.25) [Pa]

σ_e = equilibrium stress from the Figoni–Shoemaker Model (Figure 4.25) [Pa]

ϕ = particle volume fraction (Equations 4.2–4.4) [-]

A = general rheological variable (Equations 4.1, 4.6, 4.8) [$f(variable)$]

A_C = pre-exponential parameter related with the effect of concentration (Equation 4.1) [$f(variable)$]

A_T = Arrhenius's pre-exponential parameter model (Equation 4.6) [$f(variable)$]

A_{TC} = pre-exponential parameter related with the effect of temperature and concentration (Equation 4.8) [$f(variable)$]

B = factor related with the effect of concentration (Equations 4.1, 4.8) [[C]$^{-1}$]

C = product concentration (Equations 4.1, 4.5, 4.8) [°Brix, %,—]

E_a = activation energy in the Arrhenius Model (Equations 4.6 and 4.8) [J·mol^{-1}]

G' = storage modulus [Pa]

G'' = loss modulus [Pa]

k = consistency coefficient from Herschel–Bulkley Model (Figures 4.4, 4.7, 4.12B, 4.13A, 4.19, 4.25, 4.26, 4.27, 4.28A) [Pa·s^n]

k = consistency coefficient from Ostwald–de Waele Model (Figures 4.8, 4.10, 4.13B, 4.21, 4.22) [Pa·s^n]

k', k'' = consistency coefficient in power law model of viscoelasticity properties (Figure 4.20) [Pa·s$^{n'}$, Pa·s$^{n''}$]

k_{FS} = kinetic parameter in the Figoni–Shoemaker Model (Figure 4.25) [s^{-1}]

n = flow behavior index, from Herschel–Bulkley Model (Figures 4.1, 4.4, 4.7, 4.13A, 4.19, 4.25, 4.26, 4.27, 4.28A) [-]

n = flow behavior index, from Ostwald–de Waele Model (Figure 4.8, 4.13B, 4.22) [-]

n', n'' = behavior index in power law model of viscoelasticity properties (Figures 4.9A, 4.20) [-]

R = constant of the ideal gases (Equations 4.6 and 4.8) [= 8.314 Pa·m^3·mol^{-1}·K^{-1}]

T = absolute temperature (Equations 4.6 and 4.8) [K]

References

Ahmed, J. and Ramaswamy, H. S. 2006. Viscoelastic properties of sweet potato puree infant food. *Journal of Food Engineering*, 74, 376–382.

Augusto, P. E. D., Cristianini, M. and Ibarz, A. 2012a. Effect of temperature on dynamic and steady-state shear rheological properties of siriguela (Spondias purpurea L.) pulp. *Journal of Food Engineering*, 108, 283–289.

Augusto, P. E. D., Falguera, V., Cristianini, M. and Ibarz, A. 2011. Influence of fibre addition on the rheological properties of peach juice. *International Journal of Food Science and Technology*, 46, 1086–1092.

Augusto, P. E. D., Ibarz, A. and Cristianini, M. 2012b. Effect of high pressure homogenization (HPH) on the rheological properties of tomato juice: Time-dependent and steady-state shear. *Journal of Food Engineering*, 111, 570–579.

Augusto, P. E. D., Ibarz, A. and Cristianini, M. 2012c. Effect of high pressure homogenization (HPH) on the rheological properties of tomato juice: time-dependent and steady-state shear. *Journal of Food Engineering*, 111, 570–579.

Benoit, S. M., Afizah, M. N., Ruttarattanamongkol, K. and Rizvi, S. S. H. 2013. Effect of pH and temperature on the viscosity of texturized and commercial whey protein dispersions. *International Journal of Food Properties*, 16, 322–330.

Bi, X., Hemar, Y., Balaban, M. O. and Liao, X. 2015. The effect of ultrasound on particle size, color, viscosity and polyphenol oxidase activity of diluted avocado puree. *Ultrasonics Sonochemistry*, 27, 567–575.

Córdoba, A., Del Mar Camacho, M. and Martínez-Navarrete, N. 2012. Rheological behaviour of an insoluble lemon fibre as affected by stirring, temperature, time and storage. *Food and Bioprocess Technology*, 5, 1083–1092.

Dutta, D., Dutta, A., Raychaudhuri, U. and Chakraborty, R. 2006. Rheological characteristics and thermal degradation kinetics of beta-carotene in pumpkin puree. *Journal of Food Engineering*, 76, 538–546.

Genovese, D. B., Lozano, J. E. and Rao, M. A. 2007. The rheology of colloidal and noncolloidal food dispersions. *Journal of Food Science*, 72, R11–R20.

Giura, L., Urtasun, L., Ansorena, D. and Astiasarán, I. 2022. Effect of freezing on the rheological characteristics of protein enriched vegetable puree containing different hydrocolloids for dysphagia diets. *LWT*, 169, 114029.

Hsu, K.-C. 2008. Evaluation of processing qualities of tomato juice induced by thermal and pressure processing. *LWT—Food Science and Technology*, 41, 450–459.

Ibarz, A. and Barbosa-Cánovas, G. V. 2002. *Unit operations in food engineering*, CRC Press.

Ibarz, R., Falguera, V., Garvín, A., Garza, S., Pagán, J. and Ibarz, A. 2009. Flow behavior of clarified orange juice at low temperatures. *Journal of Texture Studies*, 40, 445–456.

Ibrahim, G. E., Hassan, I. M., Abd-Elrashid, A. M., El-Massry, K. F., Eh-Ghorab, A. H., Manal, M. R. and Osman, F. 2011. Effect of clouding agents on the quality of apple juice during storage. *Food Hydrocolloids*, 25, 91–97.

Khouryieh, H. A., Herald, T. J., Aramouni, F. and Alavi, S. 2007. Intrinsic viscosity and viscoelastic properties of xanthan/guar mixtures in dilute solutions: effect of salt concentration on the polymer interactions. *Food Research International*, 40, 883–893.

Koocheki, A., Taherian, A. R. and Bostan, A. 2013. Studies on the steady shear flow behavior and functional properties of Lepidium perfoliatum seed gum. *Food Research International*, 50, 446–456.

Leite, T. S., Augusto, P. E. D. and Cristianini, M. 2015. Using high pressure homogenization (HPH) to change the physical properties of cashew apple juice. *Food Biophysics*, 10, 169–180.

Liaotrakoon, W., De Clercq, N., Van Hoed, V., Van De Walle, D., Lewille, B. and Dewettinck, K. 2013. Impact of thermal treatment on physicochemical, antioxidative and rheological properties of white-flesh and red-flesh dragon fruit (Hylocereus spp.) purees. *Food and Bioprocess Technology*, 6, 416–430.

Lozano, J. E. and Ibarz, A. 1994. Thixotropic behaviour of concentrated fruit pulps. *LWT—Food Science and Technology*, 27, 16–18.

Manzoor, M. F., Ahmad, N., Aadil, R. M., Rahaman, A., Ahmed, Z., Rehman, A., Siddeeg, A., Zeng, X.-A. and Manzoor, A. 2019. Impact of pulsed electric field on rheological, structural, and physicochemical properties of almond milk. *Journal of Food Process Engineering*, 42, e13299.

Massa, A., Gonzalez, C., Maestro, A., Labanda, J. and Ibarz, A. 2010. Rheological characterization of peach purees. *Journal of Texture Studies*, 41, 532–548.

Metzner, A. 1985. Rheology of suspensions in polymeric liquids. *Journal of Rheology*, 29, 739–775.

Opazo-Navarrete, M., Tabilo-Munizaga, G., Vega-Gálvez, A., Miranda, M. and Pérez-Won, M. 2012. Effects of high hydrostatic pressure (HHP) on the rheological properties of Aloe vera suspensions (Aloe barbadensis Miller). *Innovative Food Science & Emerging Technologies*, 16, 243–250.

Osorio, M. N., Moyano, D. F., Murillo, W., Murillo, E., Ibarz, A. and Solanilla, J. F. 2017. Functional and rheological properties of piñuela (Bromelia karatas) in two ripening stages. *International Journal of Food Engineering*, 13.

Pereira, C. G., De Resende, J. V. and Giarola, T. M. O. 2014. Relationship between the thermal conductivity and rheological behavior of acerola pulp: effect of concentration and temperature. *LWT—Food Science and Technology*, 58, 446–453.

Rao, M., Cooley, H. and Vitali, A. 1984. Flow properties of concentrated juices at low temperatures. *Food Technology*, 38, 113–119.

Rao, M. A. 2014. Rheological properties of fluid foods. In *Engineering properties of food* (Rao, M. A., Rizvi, S. S., Datta, A. K. & Ahmed, J., eds.), 4th ed., CRC Press.

Rao, M. A., Cooley, H. J. and Liao, H. J. 1999. High temperature rheology of tomato puree and starch dispersion with a direct-drive viscometer. *Journal of Food Process Engineering*, 22, 29–40.

Rojas, M. L., Leite, T. S., Cristianini, M., Alvim, I. D. and Augusto, P. E. D. 2016. Peach juice processed by the ultrasound technology: Changes in its microstructure improve its physical properties and stability. *Food Research International*, 82, 22–33.

Ros-Polski, V., Schmidt, F. L., Vitali, A. A., Marsaioli J. R., A. and Raghavan, V. G. S. 2014. Rheological analysis of sucrose solution at high temperatures using a microwave-heated pressurized capillary rheometer. *Journal of Food Science*, 79, E540–E545.

Salminen, H. and Weiss, J. 2014. Effect of pectin type on association and pH stability of whey protein—pectin complexes. *Food Biophysics*, 9, 29–38.

Sato, A. C. K. and Cunha, R. L. 2009. Effect of particle size on rheological properties of jaboticaba pulp. *Journal of Food Engineering*, 91, 566–570.

Servais, C., Jones, R. and Roberts, I. 2002. The influence of particle size distribution on the processing of food. *Journal of Food Engineering*, 51, 201–208.

Shamsudin, R., Wan Daud, W. R., Takrif, M. S., Hassan, O. and Ilicali, C. 2009. Rheological properties of josapine pineapple juice at different stages of maturity. *International Journal of Food Science & Technology*, 44, 757–762.

Sharma, M., Kristo, E., Corredig, M. and Duizer, L. 2017. Effect of hydrocolloid type on texture of pureed carrots: rheological and sensory measures. *Food Hydrocolloids*, 63, 478–487.

Soria, A. C. and Villamiel, M. 2010. Effect of ultrasound on the technological properties and bioactivity of food: a review. *Trends in Food Science & Technology*, 21, 323–331.

Syahariza, Z. A. and Yong, H. Y. 2017. Evaluation of rheological and textural properties of texture-modified rice porridge using tapioca and sago starch as thickener. *Journal of Food Measurement and Characterization*, 11, 1586–1591.

Takada, N. and Nelson, P. E. 1983. Pectin-protein interaction in tomato products. *Journal of Food Science*, 48, 1408–1411.

Tavares, D. T., Alcantara, M. R., Tadini, C. C. and Telis-Romero, J. 2007. Rheological properties of frozen concentrated orange juice (FCOJ) as a function of concentration and subzero temperatures. *International Journal of Food Properties*, 10, 829–839.

Tsai, S. C. and Zammouri, K. 1988. Role of interparticular Van der Waals force in rheology of concentrated suspensions. *Journal of Rheology*, 32, 737–750.

Varela, P., Salvador, A. and Fiszman, S. 2007. Changes in apple tissue with storage time: Rheological, textural and microstructural analyses. *Journal of Food Engineering*, 78, 622–629.

Wang, J., Wang, J., Vanga, S. K. and Raghavan, V. 2020. High-intensity ultrasound processing of kiwifruit juice: Effects on the microstructure, pectin, carbohydrates and rheological properties. *Food Chemistry*, 313, 126121.

Wei, Y. P., Wang, C. S. and Wu, J. S. B. 2001. Flow properties of fruit fillings. *Food Research International*, 34, 377–381.

Yoo, B. and Rao, M. A. 1994. Effect of unimodal particle size and pulp content on rheological properties of tomato puree. *Journal of Texture Studies*, 25, 421–436.

Zhu, D., Shen, Y., Wei, L., Xu, L., Cao, X., Liu, H. and Li, J. 2020. Effect of particle size on the stability and flavor of cloudy apple juice. *Food Chemistry*, 328, 126967.

Assessing Rheological Properties of Fluid and Semi-Solid Food Products
Rheometry

Pedro E. D. Augusto, Alberto C. Miano and Meliza L. Rojas

5.1 Introduction

The main rheological procedures are based on fundamental methods using rheometers (also called rotational viscometers) with different operating procedures and geometries, pressure-driven flow viscometers (as the tube viscometers) and extensional flow viscometers (Ibarz and Barbosa-Canovas, 2014; Rao, 2013). By using fundamental procedures, a mathematical description of the rheological phenomenon is obtained. Therefore, the product properties can be suitably described, as well as its behavior during processing, storing and consumption.

However, other empirical procedures and equipment are often used in the food industry, such as the Adams consistometer and the Bostwick consistometer, as well as procedures whose design only allows one-condition experiments, such as the falling ball viscometer and the glass capillary viscometer. In these situations, the mathematical analysis is difficult or impossible, and the obtained information can be poor and inadequate for further analysis (although it can be reasonable for quality control purposes).

Rheometers are the most versatile and complete equipment which can be used to determine all the rheological properties of foods (such as the steady-state shear, time-dependent and viscoelastic properties).

In using a rheometer, the sample is placed between a moving geometry, which is connected to a motor to be rotationally moved, and a stationary base (Figure 5.1). One can imagine this procedure is similar to the imaginary experiment described in Chapter 2 (Figure 2.1), whereby the fluid is placed between a moving plate and a stationary plate. Therefore, by moving the geometry, the product in contact with the geometry surface

DOI: 10.1201/9781003148722-5

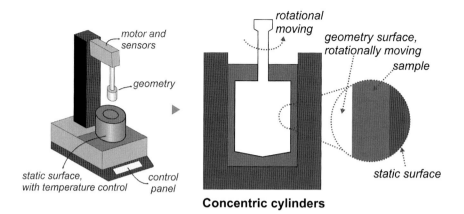

Concentric cylinders

Figure 5.1 *General view of a rheometer, with its main parts—geometry and the static surface, with temperature control. Different geometries are shown in Figure 5.2.*

will follow it, rotating with the same velocity. Similarly, the product in contact with the static surface will stay at rest. Therefore, a gradient of velocity will take place across the sample, and the rheological parameters could then be calculated and evaluated. In fact, the geometry can be rotationally moved in a continuous or oscillating way, depending on the desired purpose, resulting in different possible protocols.

It is important to highlight that food products are very complex materials, as well as their rheological behavior. Therefore, the rheological procedures must be carried out with extreme care, as many sources of errors can compromise the obtained results. For instance, some properties must be measured without any sample disturbance, with the intact structure, while some others must ensure a specific sample modification until a steady-state condition is achieved. Moreover, some assay conditions that are adequate for one sample are inadequate for another—leading to errors resulting from turbulence or sample modification, for example, as described later in this chapter. Therefore, the technique, sample, equipment and accessories limitations must be known and evaluated—in fact, it is common for the specialist to be unable to measure some properties of the evaluated material due to limitations of one or more of these issues.

In simple words, *to perform a rheological evaluation is not to follow a recipe!* It means that, contrary to some other procedures conducted for food evaluation, whereby there is a fixed protocol to be followed (an example is some chemical evaluation, many times standardized by an institution), the exact conditions to perform a rheological evaluation must be designed, tested and evaluated for each case (i.e., sample structure and properties, the desired information, considering the reason for making that evaluation). This chapter describes the main procedures used to evaluate food products and provides some tips to help the reader to better understand and perform the procedures, as well as to avoid the main observed errors. If you understand the principles of each procedure and the information you expected from your sample, you will be able to design your own protocol, extracting the best result from each assay.

5.2 General Aspects: Rheometer, Procedures and Geometries

Figure 5.1 shows a general view of a rheometer, whereby the sample is placed between a rotationally moving geometry and a stationary basis, being submitted to a continuous or oscillatory movement as a function of the desired protocol.

To assess the steady-state shear properties (fluid flow) and time-dependent flow properties (Chapter 2), the geometry continuously rotates, with a controlled shear stress (σ) being varied, and thus reading, with a sensor, the correspondent shear rate ($\dot{\gamma}$)—or vice-versa. To assess the viscoelastic properties (Chapter 3), the geometry rotates in an oscillatory way, which is defined by three parameters: shear stress (σ), specific strain (γ) and oscillatory frequency (ω). To perform those experiments, one parameter must be fixed, a second must vary and the third will be read as a response—for example, in frequency sweep experiments, the shear stress (σ) within the sample linear range is fixed, the oscillatory frequency (ω) is varied, and the specific strain (γ) is read to calculate the desired rheological properties, such as the storage (G') and loss (G") moduli. However, by fixing plus parameters, the effect of other physical quantities on the product properties can be evaluated, such as time and temperature (as described later in this chapter).

In fact, it is worth remembering that temperature must be controlled during all the experiments (Chapter 4), unless it is intentionally varied, for example to evaluate gel formation or collapsing, in temperature sweep experiments, as described in what follows.

With the obtained data, a quantitative analysis can be carried out, based on the mathematical description of the physical phenomenon, in order to obtain the product rheological behavior.

To assess the time-dependent flow properties (Chapter 2), the sample structure must be intact at the beginning of the procedure, being sheared in controlled conditions until reaching the steady-state condition (and structure). The changes in the product behavior during this procedure (which are associated with the changes on product structure) are exactly what is intended to be measured.

To assess the viscoelastic properties (Chapter 3), the sample structure must be intact during all the procedures in order to assess its mechanical response to the oscillatory movement.

Therefore, for both procedures, the sample must be placed carefully in the rheometer, trying to avoid any modification of the internal structure, and the sample is left at rest for some time (typically from 5–20 min) to recover its original state and achieve the desired temperature.

On the other hand, to assess the steady-state shear properties (fluid flow, Chapter 2) of products, the sample structure must be in steady-state flow conditions. In this case, the sample must be placed in the rheometer and controlled sheared until it reaches a stationary behavior (and structure).

Therefore, a clever strategy is to combine the experiments in order to not only save time and samples but also measuring different properties with the same sample (which can be interesting for further evaluation and discussion, in special considering that food products can present high variability).

For instance, if correctly designed, you can use the same sample to perform a protocol of (time-dependent flow) + (steady-state shear, fluid flow) properties, or (viscoelastic) + (steady-state shear, fluid flow) properties, as described in what as follows. Although it is theoretically possible to perform the three assays in a row, (viscoelastic) + (time-dependent flow) + (steady-state shear, fluid flow) properties, we think it is difficult to guarantee the necessary conditions to do that, and we therefore do not recommend it. For instance, although the sample structure is supposedly kept intact during all the viscoelastic procedures, it is difficult to confirm this, which impairs the sequential time-dependent flow evaluation.

We highlight again that each procedure and parameter must be pre-evaluated for each sample before the final evaluation. The main issues to be evaluated are the kind, dimensions, material and surface of the geometry to be used; the range of shear rate, shear stress, strain, oscillatory frequency, temperature and time to be used; proper procedure to load the sample and calculating the properties; and sample reactions during the procedure (drying, separating, chemical reactions and gelling, among others).

Different geometries are available to assess the rheological properties of food; the most important ones are presented in Figure 5.2. The mathematical description of the associated rheological phenomenon can be seen in Ibarz and Barbosa-Canovas (2014) and Rao (2013), and different aspects must be considered to choose the geometry.

The first tip to selecting the geometry is to use the experience, which can be from the reader and/or from previous works in the literature

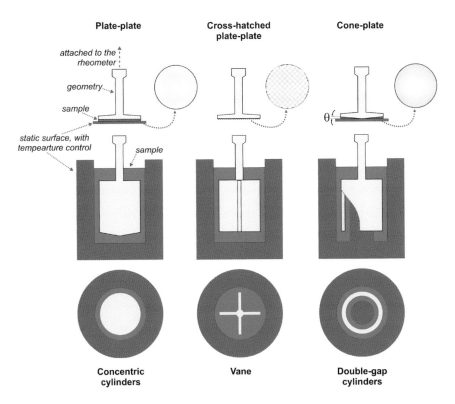

Figure 5.2 *Main geometries (in light gray) used on rheometers (rotational viscometers) to access the rheological properties of fluid and semi-solid products. The sample is colored in blue and the static surface, with temperature control, in brown.*

(with similar products and procedures). We suggest taking a look in "good works" of the literature to avoid some mistakes. Moreover, you need to understand the characteristics of your sample, which will direct the choice.

For example, in general, the less consistent the sample is, the wider is the necessary geometry area to attain the desired sensibility. Therefore, high- and medium-consistent food can be evaluated using plate-plate geometry, although the medium-consistent product can demand a longer diameter in relation to the higher consistent product. Moreover, it can be difficult to obtain reliable measurements using a plate-plate geometry for a low-consistent food, necessitating use of concentric cylinders geometry (also called Couette or bob-and-cup). In the same idea, for a very low-consistent product, the concentric cylinders geometry may not be enough, and a double gap cylinder—with both internal and external areas in contact with the product—may be needed.

Furthermore, geometry inertia can also affect the measurement. For instance, a metallic plate-plate geometry has greater inertia than an acrylic one; therefore, for a low-consistent fluid, the acrylic geometry may be better suited.

Observe that the concepts of "high," "medium," "low" and "very low" consistency/viscosity are relative—for this reason, you need to know at least the order of magnitude of your sample properties to start evaluating the sample.

Cone-plate geometry is made of a truncated cone placed over the stationary plate (surface). The cone can present different angles (θ in Figure 5.2), being in general from 1–3°. For this geometry, the mathematical derivation from the physic phenomenon is the best; thus, it is a good choice. However, in general, it is not compatible with particulate products, as with many foods, since the distance between the geometry (upper surface) and the static surface (down) are variable and the dispersed particles can be stuck between them. As illustrated in Figure 5.3, this fact can disrupt the suspended cells, irreversibly changing the sample, and/or measuring higher values of shear stress (as a result of the particle compression, not corresponding to the sample resistance to flow).

For dispersions, thus, in general, the plate-plate geometry is preferred. However, if high shear rates are employed, the fluid can slip in the surfaces, separating the suspended particles from the continuous phase (fluid). In those cases, a cross-hatched (or grooved) surface is needed, as is usual for fruit juices, derivates and other food products.

For instance, when evaluating the effect of peach fiber content in peach juice rheology, we used concentric cylinders to evaluate the steady-state

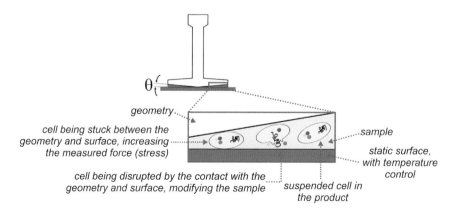

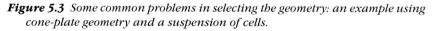

Figure 5.3 *Some common problems in selecting the geometry: an example using cone-plate geometry and a suspension of cells.*

shear properties of juices up to 10% of fiber, finding it necessary to change the geometry to a cross-hatched plate-plate when the fiber content was 12.5% (Augusto et al., 2011).

Other points must be considered when choosing the geometry, not necessarily related to the sample needs, but with some choices one may have. For instance, the required quantity of sample to run an experiment is very small using the cone-plate or plate-plate geometries (in the order of 0.5–1 mL), while it can reach values of 10–20 mL (or more) when using the concentric cylinders or vanes. On the other hand, the time and effort between two runs are higher when using the concentric cylinders or vane geometries in comparison with cone-plate or plate-plate geometries: when using concentric cylinders or vanes, the cup and geometries must be removed between samples for cleaning, which also add a necessary mapping procedure (to calibrate the geometry position in the rheometer). It is worth mentioning that the operational needs of each piece of equipment must be verified with the supplier.

Finally, the conditions of the rheological assay must be set.

Temperature must be controlled, and the normal concerns related to thermal effects on foods must be considered. For instance, by increasing temperature, the sample can experience changes in structure and composition—from protein denaturation to Maillard reaction, passing through vaporization. In fact, a gap that food rheologists and process engineers still face is how to assess the rheological properties of foods at high temperatures (such as those for ultra-high temperature [UHT], sterilization processes)—see Ros-Polski et al. (2014), for further discussion. In fact, even considering advances in equipment, including solvent traps, it is difficult to obtain reliable data in temperatures higher than 80–90°C.

The following sections describe the main aspects to assessing rheological properties (steady-state shear, time-dependent and viscoelastic properties) of fluid and semi-solid foods using rheometers.

5.3 Fluid Flow: Steady-State Shear and Time-Dependent Flow Properties

As described, the sample conditions are different to perform steady-state shear and time-dependent flow protocols. To assess the time-dependent flow properties (Chapter 2), the sample structure must be intact at the beginning of the procedure, being sheared in controlled conditions until reaching the steady-state condition (and structure)—from which the steady-state shear protocol can be conducted. Those protocols can be conducted independently or in sequence.

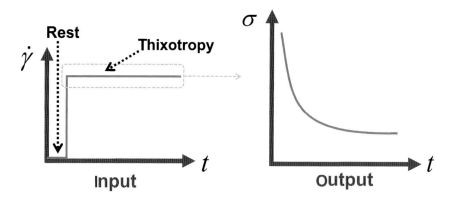

Figure 5.4 *A protocol to evaluate the time-dependent flow properties of foods.*

When the time-dependent flow protocol is conducted, the sample must be placed carefully in the rheometer, trying to avoid any modification of the internal structure, and the sample is left at rest for some time to recover its original state and achieve the desired temperature. This time must be set after pre-tests, although the order of magnitude can be selected from the literature (being typically from 5–20 minutes). After resting, the sample is sheared at a pre-selected shear rate (input of the protocol), the associated shear stress being recorded over the experimental time (output of the protocol, the results to be evaluated). This procedure is illustrated in Figure 5.4, the results having been evaluated using the models described in Chapter 2. Once this procedure is conducted with a shear rate, it can be repeated with other shear rate values (using new samples). However, as described, once the sample reaches the steady-state condition, the procedure can be enlarged to also evaluate the product flow properties—as described in what as follows.

Figure 5.5 shows two other protocols often found in the literature to evaluate the time-dependent flow and steady-state shear properties of foods. In those protocols, after the rest period, the shear rate is increased and then decreased, with the shear stress being recorded over the two ramps. If the product behavior is not time-dependent, the output curves of shear stress versus shear rate would be exactly the same. However, if those curves do not coincide, the area between them is an indicative of the product thixotropy. Although this idea is true, we see two problems with this approach. First, the product must reach the equilibrium to allow an appropriate evaluation, and it cannot be guarantee by making two ramps (up and down). Therefore, a cycle of three ramps can be conducted: up, down and up again. If the second and third curves (of shear stress versus shear rate) coincide, the sample reaches equilibrium—but if they do not coincide, the third curve represents the

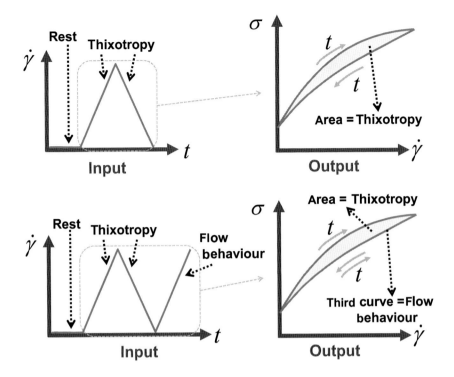

Figure 5.5 *Protocols often found in the literature to evaluate the time-dependent flow and steady-state shear properties of foods.*

steady-state conditions and can also be used to evaluate the fluid flow. Second, although the hysteresis area can be an indicative of product thixotropy, its magnitude is a function not only of the product's properties, but also the chosen parameters for the protocol (such as the maximum shear rate). Therefore, we prefer to avoid the protocols illustrated in Figure 5.5. We think that, with the same effort, one can obtain better results with other pattern of protocol (using the protocols illustrated in Figure 5.4 and Figure 5.6, and their combination in Figure 5.8).

When the steady-state shear flow protocol is conducted, the sample must be placed in the rheometer and left at rest for some time to achieve the desired temperature. Then, the sample is sheared at a pre-selected shear rate (input of the protocol; the shear rate is instantaneously applied), the associated shear stress being recorded (output of the protocol, the results to be evaluated) when steady-state condition is achieved. Then, the shear rate is stepwise decreased, in order to scan different conditions, until close to the rest (where the yield stress [σ_o] can be obtained). This procedure is illustrated in Figure 5.6, the results being evaluated using the models described in Chapter 2. Observe that the opposite curve can

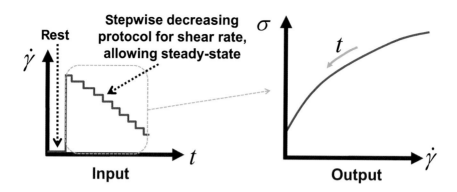

Figure 5.6 *A protocol to evaluate the steady-state shear flow properties of foods.*

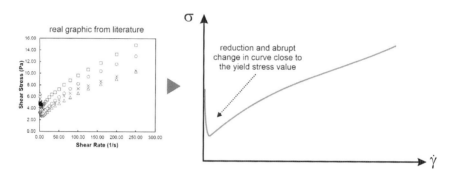

Figure 5.7 *A typical problem in selecting the conditions for the rheological protocol: by conducting a stepwise increase in shear rate protocol in fluids with yield stress (σ_0), an erroneous behavior can be seen at the beginning of the curve. The figure shows real data on fruit juices, from the literature.*

also be obtained, i.e., conducting a stepwise increasing in shear rate. However, some errors can arise by starting at rest with samples that present yield stress. Once at stress values below the yield stress, the material deforms elastically, without flowing, and an overshooting can be obtained at the beginning of the curve, as illustrated in Figure 5.7.

Finally, it is possible to combine two evaluations in a single protocol, performing a (time-dependent flow) + (steady-state shear, fluid flow) protocol, as illustrated in Figure 5.8. It can both save your time and allow more reliable results. We suggest applying this protocol, which will allow you to guarantee reaching the equilibrium in the first part (where thixotropy can be evaluated) and the steady-state condition in the second part (where the flow properties can be evaluated). In this protocol, the sample must be placed carefully in the rheometer, trying to avoid any modification of the internal structure, and the sample is left at rest

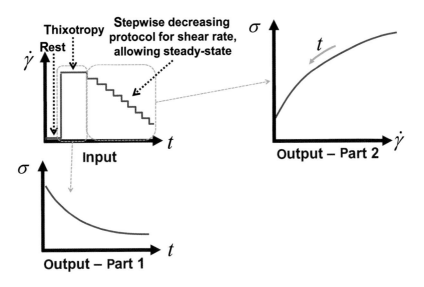

Figure 5.8 *A protocol to evaluate the time-dependent flow properties + the steady-state shear flow properties of foods, using the same sample.*

for some time to recover its original state and achieve the desired temperature. Then, the sample is sheared at a pre-selected shear rate (input of the protocol, part 1), being the associated shear stress recorded over the experimental time (output of the protocol, part 1, the results to be evaluated for thixotropy). Finally, the shear rate is stepwise decreased, in order to scan different conditions, until close to the rest (where the yield stress (σ_0) can be obtained)—part 2.

An important issue in flow protocols is to select the appropriate range of shear rate/stress in order to avoid modifications in the structure (and thus, properties) of the evaluated foods. The shear rate range is limited in this kind of procedure, being function of the equipment, geometries and samples. The lower values of shear rate are in the order of 10^{-2} s^{-1} (Rao, 2013), which can compromise the yield stress (σ_0) value determination, but that represents well most of the unit operations involved in food products processing (Figure 2.2). On the other hand, the higher values of shear rate are in the order of 10^2 s^{-1}, being function of the evaluated product. At higher shear rates, Taylor's vortices can occur (Steffe, 1996), whereby turbulence disrupts rheological analysis, or the sample can be squeezed from the geometry.

For instance, the rheological protocol and calculi are based in the fact that the fluid flow is laminar. By increasing the shear rate in a steady-state shear protocol, however, the product flow can exceed the laminar regime, impairing the evaluation. Figure 5.9 shows real data from

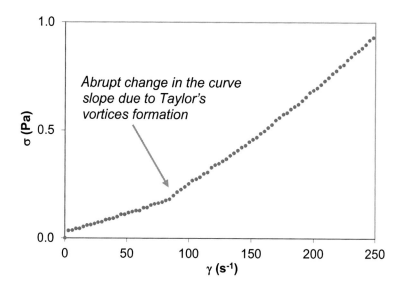

Figure 5.9 *A typical problem in selecting the conditions for the rheological protocol: by increasing the shear rate in a steady-state shear protocol, the product flow can exceed the laminar regime, impairing the evaluation. The figure shows real data from clarified peach juice (a low-viscosity Newtonian fluid).*

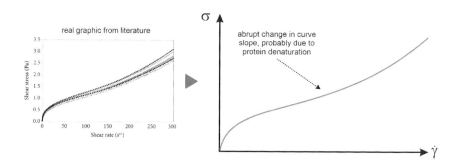

Figure 5.10 *A typical problem in selecting the conditions for the rheological protocol: by increasing the shear rate in a steady-state shear protocol, the associated shear stress can change the product structure, such as by protein denaturation. The figure shows real data from the literature, in protein-added juice.*

clarified peach juice (a low-viscosity Newtonian fluid), where Taylor's vortices are formed in conditions higher than ~80 s^{-1}, limiting the evaluation to this range. Another example is shown in Figure 5.10 for a protein-added juice: by increasing the shear rate in a steady-state shear protocol, the associated shear stress promoted protein denaturation, changing the product (which can be seen by an abrupt change in the curve slope).

Once the rheological procedure is carried out, the rheological model parameters (Chapter 2) are determined by linear or non-linear regression to describe the product properties.

5.4 Viscoelastic Properties

To assess the viscoelastic properties (Chapter 3), the sample structure must be intact during all the procedures in order to assess its mechanical response to the oscillatory movement.

The dynamic oscillatory shear procedure is conducted by imposing small-amplitude (< 5%—Gunasekaran and Ak, [2000]) oscillatory movement to the sample. The main oscillatory procedure used to describe the food product properties is the frequency sweep, whereby the strain (γ) or the shear stress (σ) is kept constant, and its response is evaluated as a function of the oscillatory frequency (ω)—these being the three variables within the product linear viscoelastic range. In this experiment, a sinusoidal oscillating movement is applied to the material, and the phase difference between the oscillating stress and strain is measured (Rao, 2013). By using the relations detailed in Chapter 3, different rheological properties (G', G'', G*, η*) can be obtained.

To perform these experiments, one parameter must be fixed, the other varied and the third read as a response. For example, in frequency sweep protocols used to assess the solid and liquid behavior of foods (through the storage [G'] and loss [G''] moduli; Chapter 3), the shear stress (σ) is fixed within the sample linear range (LVR, typical in the order of 0.1–10 Pa). This means that a previous experiment (through a stress sweep protocol) is necessary to determine the LVR, and thus the shear stress to be applied (input of the protocol, fixed). Then, the oscillatory frequency (ω) is varied (typically from 0.1 Hz to 10–100 Hz), and the specific strain (γ) is read to calculate the desired rheological properties (G', G'', G*, η*).

Figure 5.11 illustrate how to perform the shear stress sweep procedure to determine the LVR for further steps. After placing the sample carefully in the rheometer and waiting an appropriate length of time to recover its original state and achieve the desired temperature, a shear stress sweep is performed (typically between 0.01 and 100 Pa). In this case, the oscillatory frequency (ω) is kept fixed (in general close to 1 Hz), being the sample-specific strain (γ) recorded to calculate the desired rheological properties (G', G'', G*, η*). It can be observed from Figure 5.11 and Figure 5.18 that the sample presents a plateau for the value of G' within the LVR. The selected shear stress (σ), then, is selected for further steps (such as a frequency sweep protocol).

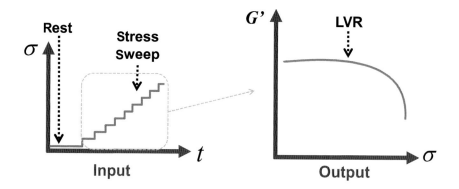

Figure 5.11 *A protocol to evaluate the linear viscoelastic range (LVR) of foods through a shear stress sweep protocol. In this protocol, the oscillatory frequency (ω) and temperature are kept constant.*

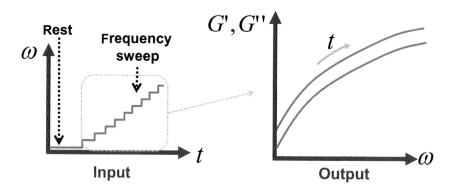

Figure 5.12 *A protocol to evaluate the viscoelastic properties of foods through a frequency sweep protocol. Further properties can be obtained from G' and G", as described in Chapter 3. In this protocol, the shear stress (σ) and temperature are kept constant.*

Figure 5.12 illustrates the protocol for a frequency sweep assay. The sample must be placed carefully in the rheometer, trying to avoid any modification of the internal structure, and the sample is left at rest for some time (typically from 5–20 minutes) to recover its original state and achieve the desired temperature, then the geometry rotates in an oscillatory manner or a specific condition is applied to the product to observe its response. As described, three parameters are related

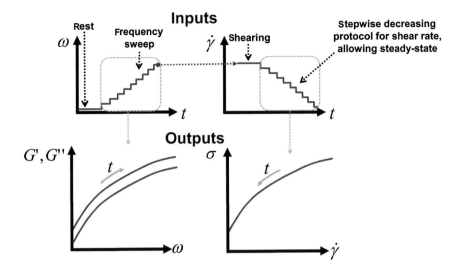

Figure 5.13 *A protocol to evaluate the viscoelastic properties through a frequency sweep protocol + the steady-state shear flow properties of foods, using the same sample.*

to this protocol: shear stress (σ), specific strain (γ) and oscillatory frequency (ω).

Time can be saved and better results obtained by performing a protocol of (viscoelastic) + (steady-state shear, fluid flow) properties with the same sample, as illustrated in Figure 5.13. In this protocol, the sample must be placed carefully in the rheometer, trying to avoid any modification of the internal structure, and the sample is left at rest for some time to recover its original state and achieve the desired temperature. Then, the oscillatory frequency (ω) is varied (typically from 0.1 Hz to 10–100 Hz), and the specific strain (γ) is read to calculate the desired rheological properties (G', G'', G*, η*), keeping constant the shear stress (σ) within the LVR. In the sequence, the sample is sheared at a pre-selected shear rate for time enough to guarantee a steady state. Finally, the shear rate is stepwise decreased in order to scan different conditions until close to the rest (where the yield stress [σ_0] can be obtained).

However, the effect of other physical quantities on the product properties can be evaluated, such as time and temperature. For instance, Figure 5.14 shows a temperature sweep protocol. Upon reaching rest, the sample is continuously heated in this protocol then oscillated at fixed oscillatory frequency (ω) and shear stress (σ), evaluating the resulting specific strain (γ) over time, in order to obtain the desired

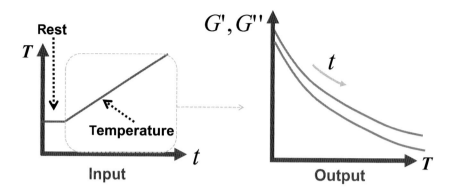

Figure 5.14 *A protocol to evaluate thermal gelling and melting through temperature sweep procedure. In this protocol, the shear stress (ω) and oscillatory frequency (σ) are kept constant. In the example of output, the sample is melting with temperature increase (as evidenced by the reduction in G' and G''), but other behaviors are possible (Chapter 4 presents some of them).*

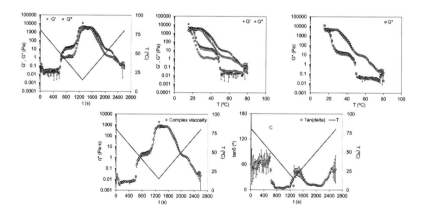

Figure 5.15 *Actual results from a protocol to evaluate gelling (by cooling) and melting (by heating) of a food gel, obtained through temperature sweep. Different results are shown. Observe how some data present a relatively high standard deviation, in special when the sample is liquid (which evidences the difficulty to perform experiments evaluating the sample when liquid and solid with the same experimental conditions).*

rheological properties (G', G'', G*, η^*). Instead of heating, the sample can be cooled, or both cooling and heating can be conducted. This protocol is interesting to evaluate gelation, melting and other phenomena related to temperature changes in food products. Figure 5.15 presents actual results from a gel, evaluating its thermal gelling (by cooling) and melting (by heating).

Similarly, time sweep protocols can be conducted to evaluate product changes over time. For instance, we conducted this protocol to evaluate the milk fermentation (Oliveira et al., 2014). The inoculated milk was placed in a glass jar, which was coupled to the rheometer cup to control temperature, using vane geometry (Figure 5.16). The vane was gently introduced into the milk, and the system was kept at the temperature desired for fermentation. The strain and oscillatory frequency values were within the minimum detectable sensitivity of the equipment to minimize interference in the gel formation, and the kinetics of gel formation were evaluated as the evolution in G' (thus, the sample solid behavior) over fermentation time. By separately measuring the change in pH over time and connecting both results with the knowledge of milk structure and physical chemistry, the different mechanisms associated with the milk gel formation—changes on electrostatic repulsion due to pH, solubilization of colloidal calcium phosphate, protein micelle weakening and precipitation, interaction among micelles—were discussed. This protocol is an interesting application of food rheology knowledge to understand the process–structure–properties relationships in food products.

Following the same idea, other protocols can be assembled to perform creep–recovery or stress–relaxation procedures. Those procedures are valuable tools to characterize the food products behavior and to correlate with structure, the description of those evaluations which is presented in Chapter 3. Contrary to oscillatory assays (frequency, temperature, stress or time sweep, whereby the geometry is moved in oscillation in both senses) and flow assays (steady-state shear and time-dependent

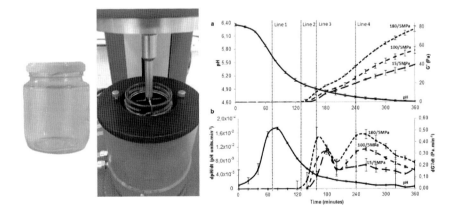

Figure 5.16 *The system used to track milk fermentation and the obtained results, demonstrating the effect of dynamic high pressure on milk fermentation kinetics and rheological properties of probiotic fermented milk.*

Source: Graphic adapted from Oliveira et al. (2014)

assays, whereby the geometry is continuously moved in one sense), the creep–recovery and stress–relaxation procedures are characterized by small movements of the geometry in one sense.

This means that a previous experiment (through a stress sweep protocol) is necessary to determine the LVR, and thus the shear stress to be applied (input of the protocol is fixed). Then, the sample is gently placed in the rheometer and, after reaching rest, the procedures are conducted, each one with two periods.

In the creep–recovery evaluation, a fixed shear stress (σ) within the LVR is imposed to the sample in the first period, being recorded its specific strain (γ) over time. Then, the stress is removed (i.e., set to zero) and the sample recovery is evaluated (i.e., its specific strain [γ] over time). The compliance (*J[t]*—Chapter 3) is then calculated for evaluation.

The stress–relaxation evaluation is the contrary. In the first period, a fixed specific strain (γ) within the LVR is imposed to the sample, being then recorded the consequent shear stress (σ) over time.

Figure 5.17 illustrates protocols to perform both creep–recovery and stress–relaxation procedures to evaluate food properties. Following the same discussion of Figure 5.13, a steady-state shear (flow) experiment can be conducted in the sequence.

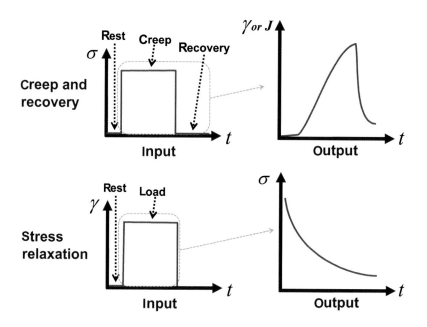

Figure 5.17 Protocols to perform creep–recovery and stress–relaxation procedures to evaluate food properties.

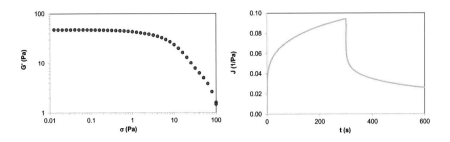

Figure 5.18 *Creep–recovery evaluation of a vegetable baby food. In the left, the shear stress sweep procedure to determine the linear viscoelastic range (LVR). In this case, the oscillatory frequency (ω) was fixed in 1 Hz, being the shear stress varied from 0.01 to 100 Pa. The LVR was set close to 1 Pa—value used in the second protocol. In the right, the second protocol performed the creep (0–250 s) and recovery (250–600 s) assays. The shear stress was kept at 1 Pa during creep, being set to 0 Pa during recovery, and the sample specific strain (γ) was recorded over time, then calculating the compliance (J(t)—Chapter 3). All the experiments were conducted at 25°C.*

Nomenclature

γ = strain [-]
$\dot{\gamma}$ = shear rate [s^{-1}]
η = viscosity [Pa·s]
η_a = apparent viscosity [Pa·s]
η_p = plastic viscosity in the Bingham Model [Pa·s]
η^* = complex viscosity [Pa·s]
σ = shear stress [Pa]
σ_0 = yield stress, Herschel–Bulkley and Bingham models [Pa]
σ_0 = initial stress in the Figoni–Shoemaker Model [Pa]
σ_e = equilibrium stress in the Figoni–Shoemaker, Hahn–Ree-Eyring and Peleg models [Pa]
ω = oscillatory frequency [Hz]
G' = storage modulus [Pa]
G'' = loss modulus [Pa]
G^* = complex modulus [Pa]
J = compliance [Pa^{-1}]
k = consistency coefficient, Herschel–Bulkley and Ostwald–de Waele models [Pa·s^n]
k', k'' = consistency coefficient in power law model of viscoelasticity [Pa·s$^{n'}$, Pa·s$^{n''}$]
n = flow behavior index, Herschel–Bulkley and Ostwald–de Waele models [-]
n', n'' = behavior index in power law model of viscoelasticity properties [-]
t = time [s]

References

Augusto, P. E. D., Falguera, V., Cristianini, M. and Ibarz, A. 2011. Influence of fibre addition on the rheological properties of peach juice. *International Journal of Food Science & Technology*, 46(5), 1086–1092. https://doi.org/10.1111/j.1365-2621.2011.02593.x

Gunasekaran, S. and Ak, M. M. 2000. Dynamic oscillatory shear testing of foods—selected applications. *Trends in Food Science & Technology*, 11(3), 115–127. https://doi.org/10.1016/S0924-2244(00)00058-3

Ibarz, A. and Barbosa-Canovas, G. V. 2014. *Introduction to food process engineering*, Taylor & Francis.

Oliveira, M. M. D., Augusto, P. E. D., Cruz, A. G. D., & Cristianini, M. 2014. Effect of dynamic high pressure on milk fermentation kinetics and rheological properties of probiotic fermented milk. *Innovative Food Science & Emerging Technologies*, 26, 67–75. https://doi.org/10.1016/j.ifset.2014.05.013

Rao, M. A. 2013. *Rheology of fluid, semisolid, and solid foods: principles and applications*, Springer Science & Business Media.

Ros-Polski, V., Schmidt, F. L., Vitali, A. A., Marsaioli Jr., A. and Raghavan, V. G. S. 2014. Rheological analysis of sucrose solution at high temperatures using a microwave-heated pressurized capillary rheometer. *Journal of Food Science*, 79(4), E540–E545. https://doi.org/10.1111/1750-3841.12398

Steffe, J. F. 1996. *Rheological methods in food process engineering*, 2nd ed., Freeman Press.

Assessing Rheological Properties of Solid Food Products
Rheometry

Alberto C. Miano, Pedro E. D. Augusto and Meliza L. Rojas

6.1 Assessing Viscoelastic Properties

As it was described in Chapter 3, solid food products commonly present viscoelastic properties, meaning that they have partly viscous and partly elastic behavior. For studying the viscoelastic properties of solid food, a texture analyzer can be used, or any mechanical testing machine. This device is used to perform texture analyses of food samples, controlling some variables as displacement (strain), force (stress) and time. A conventional texture analyzer is shown in Figure 6.1, which is consisted of a loadcell, a probe, a sample place and a control panel. The load cell converts force into a measurable electric output to be displayed in the control panel. There is a specific load cell depending on the minimum force that will be analyzed. For instance, there are load cells which work at a maximum of 5, 10 or 50 kgF.

There are some tests which can be used to describe the viscoelastic properties of solid food materials. The most common tests are the stress–relaxation and creep–recovery tests.

6.1.1 Stress–Relaxation Test

The stress–relaxation test is one common test to obtain viscoelastic properties of biological material associating elastic and viscous elements. This test helps to easily understand viscoelastic properties of food; it consists of deforming a sample keeping a constant strain while the related stress over time is evaluated (Rao and Steffe, 1992). For that, cylindrical samples are commonly used which can be procured using a corer of the desired diameter.

DOI: 10.1201/9781003148722-6

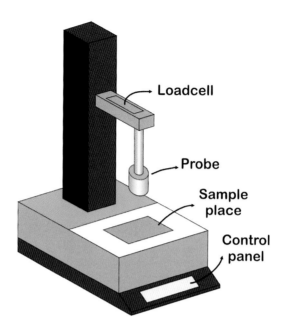

Loadcell

Probe

Sample
place

Control
panel

Figure 6.1 *General view of a texture analyzer with its main parts.*

The procedure (Figure 6.2) starts by placing the cylinder on the sampler of a texture analysis device and the sample is compressed by a cylindrical probe. It must be assured that the diameter of the probe is bigger than the sample. Furthermore, the samples should be compressed no more than 10% of their height to avoid sample fracture and significant change of diameter (2 mm for a 2 cm height cylinder, for instance). In addition, the compression velocity should be slow, suggesting 0.2–0.5 mm/s. After reaching the target strain, it is kept for some time (20–30 seconds, for instance). Finally, the compression is released.

Figure 6.3 shows the common profile (stress vs. time) of the whole stress–relaxation procedure. In addition, it is shown the compression velocity profile during the process. Notice that the probe can go down at a high velocity until reaching the sample (the equipment can automatically detect when the probe reaches the sample), after that, the velocity is reduced to compress the sample carefully until reaching the displacement distance (deformation) established in the equipment. Then the velocity of the probe is not controlled during the test, once the target is to keep the desired strain. Next, the probe can go up fast to its original place because this part of the procedure is not considered for further calculations.

The commons relaxation curves are presented in Figure 6.4. The viscoelastic behavior is represented by an upward concave shape curve

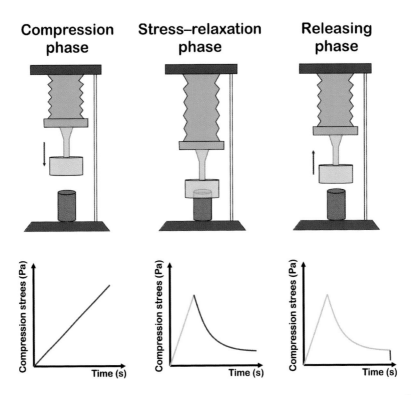

Figure 6.2 *Stress–relaxation procedure steps and its respective stress profile pattern.*

with an inclination which depends on how elastic or how viscous the material is. As the curve becomes more sloped, the material presents more viscous behavior (viscoelastic material 2). Otherwise, the material presents a more elastic behavior (viscoelastic material 1). This could be useful to qualitatively discriminate the material. Nonetheless, there are quantitative methods to determine how viscous or elastic the material behaves, being helpful for material characterization or for evaluating if the process changes the material behavior. For that, there are some mathematical models which describe the viscoelastic behavior during stress–relaxation test.

6.1.1.1 Maxwell Model

The Maxwell Model (Maxwell, 1851) describes the relaxation curve by combining the main components of mechanical viscoelasticity (Hookean spring [E] and Newtonian dashpot [η]). This model consists of a spring and a dashpot in series, which is called a Maxwell body. These bodies

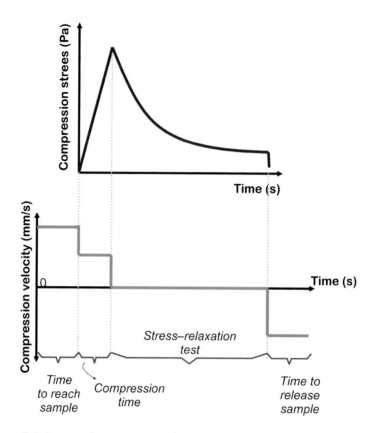

Figure 6.3 *Stress–relaxation procedure: common stress profile and compression velocity profile.*

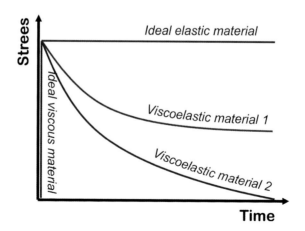

Figure 6.4 *Relaxation profiles of different kind of materials.*

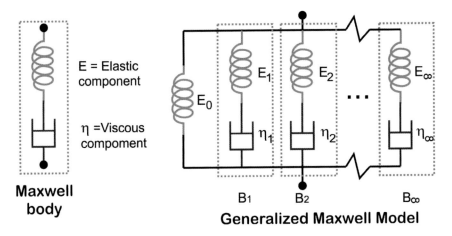

Figure 6.5 *Graphic interpretation of Generalized Maxwell Model.*

can be arranged in parallel (Figure 6.5), conforming the GM Model. This can be mathematically represented by Equation 6.1.

$$E_t = E_1 \cdot e^{-t/\tau_1} + E_2 \cdot e^{-t/\tau_2} + \ldots + E_n \cdot e^{-t/\tau_n} + E_{n+1} \tag{6.1}$$

Where E_t represents the overall sample modulus of elasticity during stress–relaxation test time (pressure units); E_1, E_2, ..., E_n represents the modulus of elasticity of the spring of each Maxwell body; E_{n+1} is the modulus of elasticity of the individual spring; t is the stress–relaxation test time; and τ_n represent the relaxation time for each body.

The texturometer commonly gives us data of force (F) against time. Therefore, knowing the transversal area of the sample (A) and the ratio between the sample height after compression (displacement distance) and the total (initial) height (dL/L), the overall modulus of elasticity during stress–relaxation test time can be determined by Equation 6.2.

$$E = \frac{F/A}{dL/L} \tag{6.2}$$

In addition, the viscosity modulus (η_n) related to the dashpot of each Maxwell boy can be determined by Equation 6.3 using the relaxation time for each body (τ_n).

$$\eta_n = \tau_n \cdot E_n \tag{6.3}$$

In fact, not all the terms of the GM Model are considered. The number of necessary model elements will depend on the type of product under study (See Chapter 3). It is recommended to use the fewest terms as possible that can fit the data from the relaxation curve (Figure 6.3). For instance, two bodies and an independent spring can be enough for modeling most of food data. In addition, an interpretation of each model parameter is desirable for explaining the mechanical properties of the samples. Therefore, the more the quantity of bodies, the more difficult to get an interpretation.

For instance, Figure 6.6 shows the relaxation curve of a yellow melon cylinder (1.5 cm diameter, 1.5 cm height) which was compressed with a cylindrical probe of 35 mm of diameter until 2 mm of depth by Miano et al. (2017). The compression was kept for 30 seconds to obtain the relaxation curve. The y-axis represents the overall modulus of elasticity which was calculated by Equation 6.2, considering the area of the cylinder base and the deformation ratio of 2:15 mm.

The experimental data was suitably fit by a Maxwell Model with two bodies and an independent spring. With this, we can know the value of the modulus of elasticity of each spring and the value of relaxation time of each body which are used to calculate viscosity with Equation 6.3.

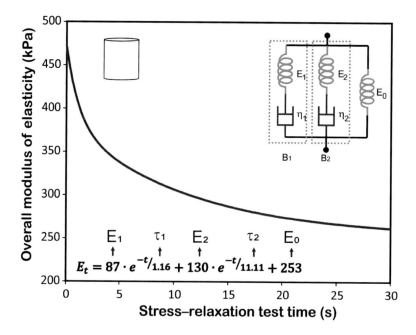

Figure 6.6 *Example of relaxation profile of yellow melon cylinders (1.5 cm diameter, 1.5 cm height). The continuous line represents the Generalized Maxwell Model which was fitted to experimental data.*

Source: Data was extracted from Miano et al. (2017).

The most difficult task is to interpret the mechanical parameters relating them to anatomical parts of the sample. For the same work with melon cylinders, the authors evaluated the structure of melon, giving a possible interpretation of the mechanical parameters of Maxwell Model related to the melon tissue. Figure 6.7 shows schematically the relation among the mechanical parameters and food structure. In that work, one Maxwell body may be related to a closed cavity (a fluid bounded by a wall), such as cells, intercellular spaces, organelles and vacuoles, among others. Regarding a cell, the fluid would be the cytoplasm, which has certain apparent viscosity related to the dashpot; on the other hand, the cell membrane would have certain elasticity, which would be related to the spring.

Furthermore, the complete model (two bodies and an independent spring) might be related to different tissue structures. One body would be related to the cells, another related to intercellular cavities and the independent spring related to free material that does not contain any fluid, such as the lamella media carbohydrates. This would be helpful for characterizing or evaluating process effects; for example, if the viscosity related to intercellular fluid increases could be related to the replacement of the internal fluid: air by water, or water by sucrose solution. If the elasticity related to the cells is reduced, this can lead to cell wall breakdown.

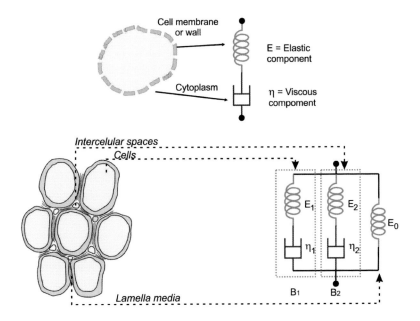

Figure 6.7 *Possible correlation between structure and mechanical bodies of Generalized Maxwell Model.*

Similar associations were made for pumpkin cylinders (Rojas and Augusto, 2018a). In addition, other works that used Maxwell Model for studying viscoelasticity of food include studies of flakes (Ozturk and Takhar, 2017) and cherries (Moghimi et al., 2011).

6.1.1.2 Peleg Model

Another mathematical model that can be used to describe viscoelasticity via stress–relaxation test is the Peleg Model (Peleg and Calzada, 1976). This is an empirical model that has two parameters (Equation 6.4): k_1 represents the reciprocal of the initial decay rate of the relaxation curve, and k_2 represents the hypothetical asymptotic level of the normalized relaxation curve. Further, the parameter k_2 has been related to "a representation of the solidity degree" (Peleg and Normand, 1983, pp. 108–113), whose value can be from 1–∞, noted as 1 for pure fluids and becoming more solid as the value approaches infinity.

For this model, the compression stress (σ) is used, which is calculated dividing the compression force by the area of application (area of the cylinder base). To obtain the normalized relaxation curve, the right term of Equation 6.4 against test time is plotted. In this case, σ_0 is the initial compression stress or its value at the beginning of the test and $\sigma(t)$ is the compression stress during the test time.

$$\frac{\sigma_0 - \sigma(t)}{\sigma_0} = \frac{t}{k_1 + k_2 \cdot t} \tag{6.4}$$

The advantage of this model is that it has only two parameters which can be used for studying any change due to processing. Having fewer parameters helps to focus attention on other variables, such as the effect of processing variables. On the other hand, the main drawback is that the parameters do not have physical meaning which can be directly related to the food structure.

Some examples of the use of this model are for evaluating apple (Rojas et al., 2020), pumpkin (Carvalho et al., 2020), dough (Lazaridou et al., 2019) and avocado (Ortiz-Viedma et al., 2018).

Regarding the work of Carvalho et al. (2020) as an example, they evidenced the reduction of parameter k_2 due to processing—which means loss of solidity, or elasticity. They used drying accelerator substances as ethanol to improve drying process of pumpkin, finding that the pretreated samples had lower value of k_2 than control samples (from 1.59 s^{-1} to 1.17 s^{-1}). This made sense, since substances as ethanol can weaken cell walls and cause turgidity loss, which reduces elasticity of the samples.

Therefore, as this example, Peleg Model can be used to study changes in the elasticity of food samples due to process.

6.1.1.3 Guo–Campanella Model

The Guo–Campanella is a model based on fractional calculus (Guo and Campanella, 2017). This model (Equation 6.5) fits the data of stress ($\sigma(t)$) as function of test time (t) and presents two parameters: k (viscoelastic modulus) and α (fractional order). In the equation, ε_0 is the constant strain (relation between deformation and total height) and Γ is the gamma function.

$$\sigma(t) = k_g \frac{\varepsilon_0}{\Gamma(1-\alpha)} t^{-\alpha} \tag{6.5}$$

When $\alpha = 0$, Equation 6.5 turns into the Hooke's Law of elasticity and k becomes the elastic modulus (Equation 6.6). When $\alpha = 1$, Equation 6.5 turns into Newton's Law of viscosity and k becomes the viscosity (Equation 6.7). Finally, when $0 < \alpha < 1$, Equation 6.5 represents the viscoelastic behavior, k being a viscoelastic modulus. This means that when the value of α tends to 0, the food behaves more like a solid, and when it tends to 1, the food behaves more like a liquid.

$$\sigma(t) = k_g \cdot \varepsilon_0 \tag{6.6}$$

$$\sigma(t) = k_g \cdot \frac{\varepsilon_0}{t} \tag{6.7}$$

This model has as a main advantage that it has only two parameters, like Peleg Model. However, the parameters have a physical meaning with can be related to the food structure. As stated before, having fewer parameters helps to better interpret the effect of food processing on food structure. This model was successfully used for potato (Guo and Campanella, 2017; Rojas and Augusto, 2018b; Miano et al., 2018a), melon, sausage, cheese and pumpkin (Augusto et al., 2017).

For example, the work of Miano et al. (2018a) evaluates the effect of ultrasound on the structure of potato cylinders. By using Guo–Campanella Model, the change on potato structure was evaluated and explained with microstructure observation. It is known that high-power ultrasound causes cell disruption, which implies changes on mechanical properties of food. Figure 6.8 shows how the value of α was increased as the ultrasound processing time increased for potato cylinders. It was evidenced that cells were disrupted causing fluids release and reducing the elasticity of the whole structure. That is why α tended to increase—however,

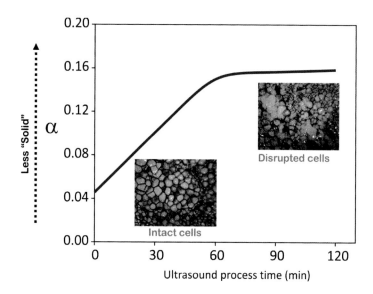

Figure 6.8 *Effect of ultrasound pre-treatment time on the fractional order of Guo–Campanella Model.*

Source: Data from Augusto et al. (2017).

with a limit, suggesting that no more significant cell disruption was caused between 60 minutes and 120 minutes of ultrasound application.

6.1.2 Creep–Recovery Test

Creep–recovery is another test to evaluate viscoelasticity of solid food. In contrast to the stress–relaxation test, this test measures the deformation of a sample due to a constant stress application across time. In this case, a sample—which can be a cylinder—is subjected to a constant force by a texture analyzer using a suitable probe such a cylindrical one. The constant force is kept for a certain time (creep phase) and then it is released (recovery phase), obtaining the profile shown in Figure 6.9.

For analyzing the results, compliance (J) against time is plotted (Figure 6.10). The compliance is calculated by Equation 6.8, the relation between strain and the applied stress, obtaining the creep compliance plot.

$$J(t) = \frac{\varepsilon(t)}{\sigma} \tag{6.8}$$

The creep compliance can be analyzed by the Burgers Model (Burgers, 1935), which is a combination in series of a Maxwell body (Hookean

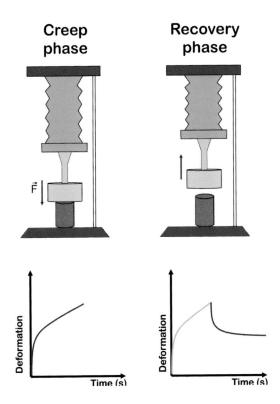

Figure 6.9 *Creep–recovery procedure steps and its respective stress profile pattern.*

spring and Newtonian dashpot arranged in series) and a Kelvin–Voigt body (Hookean spring and Newtonian dashpot arranged in parallel). Therefore, by using this model, elastic and viscous behavior can be isolated for a material. The Burgers Model is represented in Equation 6.9 in which there are parameters related to the elasticity moduli from the Maxwell body (M) and from the Kelvin–Voigt body (K). In the same way, the parameter related to viscosity (η) from the Maxwell body (M) and from the Kelvin–Voigt body (K) are also presented. In addition, Figure 6.11 represents graphically the Burgers Model, indicating which part of the curve represents each mechanical body or element. It is important to mention that more bodies can be used. For instance, a model with six parameters (one Maxwell body and two Kelvin–Voigt bodies can be used) was used to describe the viscoelasticity of a mixed ice cream (Sherman, 1966).

$$J(t) = \frac{1}{E_M} + \frac{1}{E_K}\left(1 - e^{-\frac{E_K}{\eta_K}t}\right) + \frac{1}{\eta_M} \tag{6.9}$$

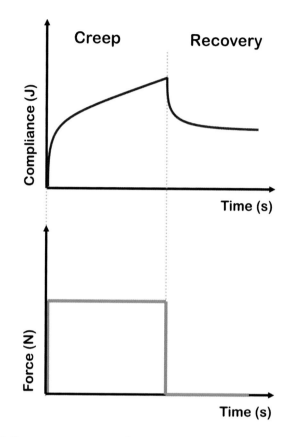

Figure 6.10 *Creep–recovery procedure: common stress profile and compression velocity profile.*

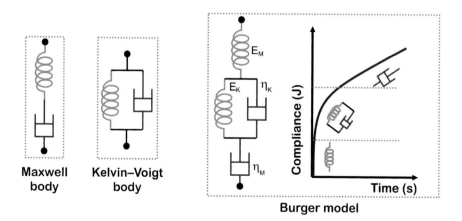

Figure 6.11 *Graphic interpretation of Burgers Model.*

Similarly to GM Model, the parameters of Burgers Model can be related to the food structure. For example, in the case of ice cream, the parameters of viscosity and elasticity are related mainly to the interaction between fat globules and ice crystals. In addition, a six-parameter model was used since fat globules less than 0.5 µm behaved differently than larger fat globules regarding molecular interaction, causing the increment of one additional Kelvin–Voigt body to the model (Dogan et al., 2013; Sherman, 1966).

Equation 6.10 can be used to analyze the recovery phase (Figure 6.9) (recovery compliance) (Dolz et al., 2008). In this model, b and c represent parameters related to the recovery speed of the system. In addition, J_K represents the recovery due to the Kelvin–Voigt body and J_∞ represents the residual deformation due to the dashpot of the Maxwell body, which is the irreversible deformation off this element (Dogan et al., 2013).

$$J(t) = J_\infty + J_K \cdot e^{-b \cdot t^c} \tag{6.10}$$

6.2 Other Mechanical Properties

Other common tests that can be used for evaluating mechanical properties of solid food are the compression test, puncture test and tensile test. These tests also use a texturometer, which is employed with the most frequent commercial probes.

6.2.1 Compression Test

This test serves to evaluate the resistance of food material to a uniaxial compression force. With this test, food materials can be characterized to look for a purpose and to evaluate the effect of any process. For instance, it can be used to evaluate gel force with different compositions, characterize fruit varieties, and evaluate effect of non-thermal processes on vegetable structures, among other applications.

The procedure is not complicated and consists of placing a sample, mostly cylindrical (similar as the stress–relaxation test) (Figure 6.12). It is crucial to assure that the cylinder diameter is shorter than the probe diameter. After placing the cylinder on the equipment's sampler, the probe should go down at a suitable velocity before touching the sample, for example at 5 mm/s. When the probe touches the sample, the velocity of the probe should be slower as 2, 1 or less than 1 mm/s to compress the sample carefully. For this test, the linear compression profile is required, so a pretest is desirable to define how deep the sample must be compressed. If the sample is compressed too much, the compression profile will be

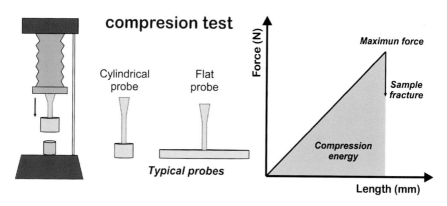

Figure 6.12 *Compression test: typically used probes and profile.*

non-linear or the sample will break down, unless the fracture force is the objective of the experiment. Finally, the probe has to go up. The velocity of the probe in the last part can be fast without any effect on the compression profile.

From the compression profile (Figure 6.12), some measurements can be taken for further analysis. The maximum force, for instance, is the last force obtained after compressing the sample to a determined depth without fracture. On the other hand, the compression energy would be the area below the profile from zero to the maximum compressed depth (force vs. distance, in appropriate units). In addition, the whole profile (linear compression, non-linear compression and fracture) can be obtained, but only considering the maximum force of the first formed peak (Cheng et al., 2019).

Examples of applications were for evaluating drying and rehydration of pineapple samples (Saavedra et al., 2022), gel force prepared by modified starch (Castanha et al., 2020) and gel force prepared by irradiated starch (Castanha et al., 2019).

6.2.2 Puncture or Penetration Test

The puncture test is similar to the compression test, but contrary to the compression test, the diameter of the probe must be much smaller than the diameter of the sample. This can be a flat-tip needle probe or a needle probe (Figure 6.13). This test is generally used for testing the coat resistance of some foods such as the epicarp (skin) of some fruits and the seed coat of grains, crusts and weak gels that cannot be demolded, among others. In addition, process effects can be also evaluated, such as the seed coat softening during grain cooking or the ripening of fruits.

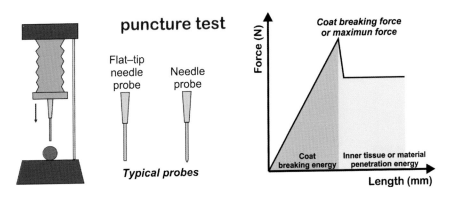

Figure 6.13 *Puncture test: typically used probes and profile.*

The procedure of this test consists of penetrating the sample with the probe until certain depth. First, the probe has to go down until reaching the sample. In this step, the velocity is not relevant. As soon as the probe touches the sample, the velocity has to be reduced for penetrating the sample, such as 0.5 mm/s. The penetration depth should be established depending on the thickness/height of the sample, taking care not to reach the sampler surface. This will assure that the coat (epicarp, seed coat, another coat) will break. Finally, the probe goes up again at an irrelevant velocity.

The puncture test creates the profile shown in Figure 6.13. First, the force linearly grows until the coat breaks, which causes a sudden force reduction due to the resistance difference between the coat and the inner tissue or material. From this step, the force will measure the resistance of the inner material of the sample which can be the cotyledon of beans, the endosperm of cereal grains, the mesocarp or endocarp of fruits or other material.

The most used value from the puncture test profile is the coat breaking force, which is the force represented by the highest peak. In addition, the energy required for breaking the coat would be the area below the curve from the beginning of penetrating until the peak of coat breaking. These values can be compared to characterized food or for evaluating process effects.

For instance, this technique can be used for evaluating the cooking process of grains, where grains are put inside boiling water and their seed coat softening is monitored by puncture testing during the processing time. This was used for such legumes as common beans (Kwofie et al., 2020), pigeon peas (Vásquez et al., 2021), mung beans (Castanha et al., 2019) and white kidney beans (Miano et al., 2018b). In addition, the puncture test has been used for studying the texture of such fruits as

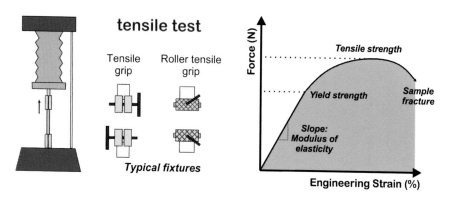

Figure 6.14 *Tensile test: typically used probes and profile.*

strawberries (Hajji et al., 2022), tomato (Kumar et al., 2022), blueberries (Rivera et al., 2022) and apples (Camps et al., 2005). Weak gels, as those obtained from cassava starch, can also be evaluated by this method by using a probe larger than a needle, but still a cylinder that penetrates the sample, such as in the work of Lima et al. (2021).

6.2.3 Tensile Test

Another conventional test is the tensile test, in which—in contrast to compression—the sample is stretched with a uniaxial force. This test can be used to evaluate such materials as plastics, bioplastics, fruit leathers, dough and cheeses, among others. In this case, specific probes are used which have grips to clamp each end of the material (Figure 6.14). Then, one grip keeps fixed while the other grip goes up, generating a tensile force. The tensile force is increased until fracture of the material is reached, ending the test. It is important to mention that the sample may have a special shape for the test, such as dumbbell/dog bone shape for some plastic material, as some standard shapes recommended by the American Society for Testing and Materials standard ASTM D638 Type I (Figure 6.15). On the other hand, the tensile test can be also performed for other sample shapes, depending on the purpose of the study, such as for noodles and cheese.

The tensile test profile is created by plotting the force against the engineering strain ($\frac{dL}{L}$). This profile provides such material information as the modulus of elasticity, which is the slope of the linear section of the profile; the yield strength, which is the maximum force where the material behaves as pure elastic; and the tensile strength, which is the maximum force of the material before it fractures.

Some examples of application of this test are for caramel (Schab et al., 2022), pumpkin rind and flesh (Shirmohammadi et al., 2013), pasta

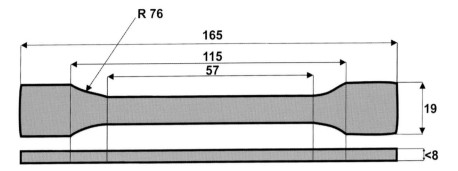

Figure 6.15 *Standard shape recommended by ASTM D638 for tensile test of materials.*

(Petitot et al., 2010) and soy protein isolate (Cunningham et al., 2000) or starch films (La Fuente et al., 2021).

Nomenclature

ε_0 = constant strain (Equation 6.5) [-]

Γ = Gamma function

η_K = apparent viscosity related to Kelvin–Voigt body (Equation 6.9) [Pa·s]

η_M = apparent viscosity related to Maxwell body (Equation 6.9) [Pa·s]

η_n = apparent viscosity related to the dashpot of the "n" Maxwell body (Equation 6.3) [Pa·s]

σ_0 = initial stress during relaxation test (Equation 6.4) [Pa]

σ_t = stress during relaxation time (Equation 6.4) [Pa]

σ_n = relaxation time of "n" Maxwell body (Equation 6.1) [Pa]

A = area (Equation 6.2) [m²]

b, c = adjustment parameters (Equation 6.10) [s⁻¹,-]

F = force (Equation 6.2) [N]

L = length of a stressed sample (Equations 6.2) [m]

k_1, k_2 = Peleg Model parameters (Equation 6.4) [s, -]

k_g = viscoelastic modulus of Guo–Campanella Model (Equation 6.5) [Pa·s$^\alpha$]

E_K = modulus of elasticity of Kelvin–Voigt body (Equation 6.9) [Pa]

E_M = modulus of elasticity of Maxwell body (Equation 6.9) [Pa]

E_n = modulus of elasticity of "n" Maxwell body (Equation 6.1) [Pa]

E_t = modulus of elasticity during relaxation time (Equation 6.1) [Pa]

J_K = compliance related to Kelvin–Voigt body recovery (Equation 6.10) [Pa⁻¹]

$J(t)$ = compliance (Equation 6.8) [Pa⁻¹]

J_∞ = compliance related to Maxwell body residual deformation (Equation 6.10) [Pa^{-1}]

t = time (Equation 6.1) [s]

References

Augusto, P. E. D., Miano, A. C. and Rojas, M. L. 2017. Evaluating the Guo-Campanella viscoelastic model. *Journal of Texture Studies*, n/a-n/a.

Burgers, J. 1935. Mechanical considerations-model systems-phenomenological theories of relaxation and of viscosity. *First Report on Viscosity and Plasticity*, 1.

Camps, C., Guillermin, P., Mauget, J. C. and Bertrand, D. 2005. Data analysis of penetrometric force/displacement curves for the characterization of whole apple fruits. *Journal of Texture Studies*, 36, 387–401.

Carvalho, G. R., Rojas, M. L., Silveira, I. and Augusto, P. E. D. 2020. Drying accelerators to enhance processing and properties: ethanol, isopropanol, acetone and acetic acid as pre-treatments to convective drying of pumpkin. *Food and Bioprocess Technology*, 13, 1984–1996.

Castanha, N., Miano, A. C., Jones, O. G., Reuhs, B. L., Campanella, O. H. and Augusto, P. E. D. 2020. Starch modification by ozone: correlating molecular structure and gel properties in different starch sources. *Food Hydrocolloids*, 108, 106027.

Castanha, N., Miano, A. C., Sabadoti, V. D. and Augusto, P. E. D. 2019. Irradiation of mung beans (Vigna radiata): a prospective study correlating the properties of starch and grains. *International Journal of Biological Macromolecules*, 129, 460–470.

Cheng, Y., Donkor, P. O., Ren, X., Wu, J., Agyemang, K., Ayim, I. and Ma, H. 2019. Effect of ultrasound pretreatment with mono-frequency and simultaneous dual frequency on the mechanical properties and microstructure of whey protein emulsion gels. *Food Hydrocolloids*, 89, 434–442.

Cunningham, P., Ogale, A., Dawson, P. and Acton, J. 2000. Tensile properties of soy protein isolate films produced by a thermal compaction technique. *Journal of Food Science*, 65, 668–671.

Dogan, M., Kayacier, A., Toker, Ö. S., Yilmaz, M. T. and Karaman, S. 2013. Steady, dynamic, creep, and recovery analysis of ice cream mixes added with different concentrations of xanthan gum. *Food and Bioprocess Technology*, 6, 1420–1433.

Dolz, M., Hernández, M. J. and Delegido, J. 2008. Creep and recovery experimental investigation of low oil content food emulsions. *Food Hydrocolloids*, 22, 421–427.

Guo, W. and Campanella, O. H. 2017. A relaxation model based on the application of fractional calculus for describing the viscoelastic behavior of potato tubers. *Transactions of the ASABE*, 60, 259–264.

Hajji, W., Gliguem, H., Bellagha, S. and Allaf, K. 2022. Structural and textural improvements of strawberry fruits by partial water removal prior to conventional freezing process. *Journal of Food Measurement and Characterization*, 16, 3344–3353.

Kumar, R., Paul, V., Pandey, R., Sahoo, R. N., Gupta, V. K., Asrey, R. and Jha, S. K. 2022. Reflectance based non-destructive assessment of tomato fruit firmness. *Plant Physiology Reports*, 27, 374–382.

Kwofie, E. M., Mba, O. I. and Ngadi, M. 2020. Classification, force deformation characteristics and cooking kinetics of common beans. *Processes*, 8, 1227.

La Fuente, C. I. A., De Souza, A. T., Tadini, C. C. and Augusto, P. E. D. 2021. A new ozonated cassava film with the addition of cellulose nanofibres: production and characterization of mechanical, barrier and functional properties. *Journal of Polymers and the Environment*, 29, 1908–1920.

Lazaridou, A., Marinopoulou, A. and Biliaderis, C. G. 2019. Impact of flour particle size and hydrothermal treatment on dough rheology and quality of barley rusks. *Food Hydrocolloids*, 87, 561–569.

Lima, D. C., Castanha, N., Maniglia, B. C., Matta Junior, M. D., La Fuente, C. I. A. and Augusto, P. E. D. 2021. Ozone processing of Cassava starch. *Ozone: Science & Engineering*, 43, 60–77.

Maxwell, J. C. 1851. On the equilibrium of elastic solids. *Proceedings of the Royal Society of Edinburgh*, 2, 294–296.

Miano, A. C., Da Costa Pereira, J., Miatelo, B. and Augusto, P. E. D. 2017. Ultrasound assisted acidification of model foods: kinetics and impact on structure and viscoelastic properties. *Food Research International*, 100, 468–476.

Miano, A. C., Rojas, M. L. and Augusto, P. E. D. 2018a. Structural changes caused by ultrasound pretreatment: direct and indirect demonstration in potato cylinders. *Ultrasonics Sonochemistry*, 52, 176–183.

Miano, A. C., Sabadoti, V. D. and Augusto, P. E. D. 2018b. Enhancing the hydration process of common beans by ultrasound and high temperatures: impact on cooking and thermodynamic properties. *Journal of Food Engineering*, 225, 53–61.

Moghimi, A., Saiedirad, M. H. and Moghadam, E. G. 2011. Interpretation of viscoelastic behaviour of sweet cherries using rheological models. *International Journal of Food Science & Technology*, 46, 855–861.

Ortiz-Viedma, J., Rodriguez, A., Vega, C., Osorio, F., Defillipi, B., Ferreira, R. and Saavedra, J. 2018. Textural, flow and viscoelastic properties of Hass avocado (Persea americana Mill.) during ripening under refrigeration conditions. *Journal of Food Engineering*, 219, 62–70.

Ozturk, O. K. and Takhar, P. S. 2017. Stress relaxation behavior of oat flakes. *Journal of Cereal Science*, 77, 84–89.

Peleg, M. and Calzada, J. F. 1976. Stress relaxation of deformed fruits and vegetables. *Journal of Food Science*, 41, 1325–1329.

Peleg, M. and Normand, M. D. 1983. Comparison of two methods for stress relaxation data presentation of solid foods. *Rheologica Acta*, 22, 108–113.

Petitot, M., Boyer, L., Minier, C. and Micard, V. 2010. Fortification of pasta with split pea and faba bean flours: pasta processing and quality evaluation. *Food Research International*, 43, 634–641.

Rao, M. A. and Steffe, J. F. 1992. *Viscoelastic properties of foods*, Elsevier Applied Science.

Rivera, S., Giongo, L., Cappai, F., Kerckhoffs, H., Sofkova-Bobcheva, S., Hutchins, D. and East, A. 2022. Blueberry firmness—a review of the textural and mechanical properties used in quality evaluations. *Postharvest Biology and Technology*, 192, 112016.

Rojas, M. L. and Augusto, P. E. 2018a. Ethanol pre-treatment improves vegetable drying and rehydration: Kinetics, mechanisms and impact on viscoelastic properties. *Journal of Food Engineering*, 233, 17–27.

Rojas, M. L. and Augusto, P. E. D. 2018b. Ethanol and ultrasound pre-treatments to improve infrared drying of potato slices. *Innovative Food Science & Emerging Technologies*, 49, 65–75.

Rojas, M. L., Augusto, P. E. D. and Cárcel, J. A. 2020. Combining ethanol pre-treatment and ultrasound-assisted drying to enhance apple chips by fortification with black carrot anthocyanin. *Journal of the Science of Food and Agriculture*, 101, 2078–2089.

Saavedra, J., De Oliveira Gomes, B., Augusto, P. E. D., Rojas, M. L. and Miano, A. C. 2022. Structure–process interaction in mass transfer processes: application of ethanol and ultrasound in a vascular structure. *Journal of Food Process Engineering*, n/a, e14187.

Schab, D., Tiedemann, L., Rohm, H. and Zahn, S. 2022. Application of a tensile test method to identify the ductile-brittle transition of caramel. *Foods*, 11, 3218.

Sherman, P. 1966. The texture of ice cream 3. Rheological properties of mix and melted ice cream. *Journal of Food Science*, 31, 707–716.

Shirmohammadi, M., Yarlagadda, P., Gu, Y. T., Gudimetla, P. and Kosse, V. 2013. Tensile properties of pumpkin peel and flesh tissue and review of current testing methods. *Transactions of the ASABE*, 56, 1521–1527.

Vásquez, U., Siche, R. and Miano, A. C. 2021. Ultrasound-assisted hydration with sodium bicarbonate solution enhances hydration-cooking of pigeon pea. *LWT*, 144, 111191.

Conclusions and Future Perspectives

Foods are complex systems in relation to their composition, structure and properties—which are connected and correlated.

In fact, food products show a complex rheology, commonly described by non-Newtonian behavior, with or without time dependency, and viscoelastic behavior. Processing changes both structure and composition, consequently affecting the food properties such as rheology. The rheological properties, in turn, dictate the product behavior during processing, storage, consumption and digestive reaction. Moreover, it is interesting to notice the rheological properties vary widely during the different steps and unit operations related to food processing, commercialization and storage. In fact, there is not any other process, condition or physical quantity that varies so much during food processing.

Therefore, it is necessary to study the rheological properties of foods to design and obtain efficient and enhanced processes and products. It seems to be a complex task, but our intention with this book was to demonstrate it is not—you just need to understand the basic concepts, where you are and where you would like to reach.

Our proposal was to provide a practical book, with a direct and didactic description of food rheology and rheometry. This book describes different food products (liquid, semi-solid, and solid; homogeneous and heterogeneous), presenting examples and applications, including our experience in aspects such as practical rheometry and main concerns, the effect of food structure and food processing through conventional and emerging technologies, among others.

We hope the book can be useful to your needs, helping you in your daily work.

DOI: 10.1201/9781003148722-7

Moreover, we would like to highlight that there are still many things to be studied and better understood in relation to food rheology.

For example, the composition–structure–processing–properties relationships still need to be further understood in food products, including rheology. New challenges, technologies and demands arise every day, summing—and not replacing—the previous ones. How can we use a specific technology to modify a food product in order to enhance its digestibility? Can we tailor food and ingredients to modulate their rheology during digestion, positively affecting nutrient absorption? How can we modulate food rheology to reduce energy consumption during processing, also achieving the best perception during consumption? Moreover, topics apparently disconnected—such as climate change, population aging and space exploration—can, in fact, present some similarities in terms of food challenges, highlighting the importance of rheology.

Climate changes not only modify the ideal places to grow and the types of food varieties that can be acclimatized (with differentiated composition and properties), but also modify how food is processed, preserved and consumed. Population aging brings challenges for medicine and even the economy, and, of course, for food products: the digestive process is changed, and the importance of texture is highlighted, from chewing to swallowing and flow. Food processing in space will become a reality (although we do not know if we will be able to see it), and real optimized processes will need to be developed; many terrestrial constraints will be discarded and food processing will need to be rethought—of course taking rheology into account.

As you can see, the fields of application for food rheology are huge. We hope you like this topic, joining us in the group of researchers committed to exploring it.

Index

Note: Numbers in **bold** indicate a table. Numbers in *italics* indicate a figure.